The understanding of circadian rhythms and the insights which they provide regarding the time-keeping ability of organisms in general have progressed remarkably within the last ten years. In writing this book, the author has endeavoured to capture the spirit of the experiments by which this was achieved, as well as the facts which those experiments brought to light. Less stress has been laid on periodic phenomena with other dimensions, since these are less well understood; and no discussion of experiments with animals has been included, although no basic concepts have been omitted because of this limitation.

The book is written for the student who wants a clear explanation of the nature and function of biological rhythms. It is intended to make it possible for him to do meaningful experiments of his own, understanding what has al-already been done, bu
bered by theoretica
It does not propose
but rather sets for
which will enable
investigator to mak
contributions.

D1219147

RHYTHMIC PHENOMENA
in
PLANTS

EXPERIMENTAL BOTANY

An International Series of Monographs

CONSULTING EDITORS

J. F. Sutcliffe

School of Biological Sciences, University of Sussex, England

AND

P. Mahlberg

Department of Botany, Indiana University, Bloomington, Indiana, U.S.A.

Volume 1. D. C. SPANNER, Introduction to Thermodynamics. 1964
Volume 2. R. O. SLATYER, Plant/Water Relationships. 1967
Volume 3. B. M. SWEENEY, Rhythmic Phenomena in Plants. 1969

IN PREPARATION

G. A. D. JACKSON, Ascorbic Acid in Plants
W. W. SCHWABE, Physiology of Flowering
N. P. KEFFORD, Regulation of Plant Growth
R. J. ELLIS, Sulphur in Plants
J. F. LONERAGEN, Trace Elements in Plants

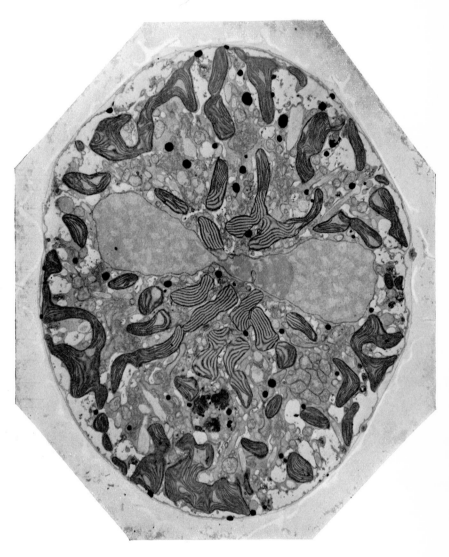

Where is the clock? *Gonyaulax polyedra* in longitudinal section showing the nucleus, chloroplasts, mitochondria and other structures.

Frontispiece

RHYTHMIC PHENOMENA
in
PLANTS

B. M. SWEENEY

Department of Biological Sciences
University of California, Santa Barbara
California, U.S.A.

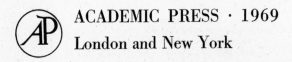

ACADEMIC PRESS · 1969
London and New York

ACADEMIC PRESS INC. (LONDON) LTD
Berkeley Square House
Berkeley Square
London, W.1X 6BA

U.S. Edition published by
ACADEMIC PRESS INC.
111 Fifth Avenue
New York, New York 10003

Library of Congress Catalog Card Number: 68–57749

PRINTED IN GREAT BRITAIN AT
BUTLER AND TANNER LIMITED
FROME AND LONDON

To the students of the future
who will reveal the balance wheel
of the biological clock

Preface

Our understanding of circadian rhythms, and the insights that they provide regarding the time-keeping ability of organisms in general, have progressed remarkably within the last ten years. In writing this book, I have tried to capture the excitement of the experiments by which this was achieved, as well as the facts which the experiments brought to light. In doing so perhaps I have neglected periodic phenomena with other dimensions. This has been in part deliberate since they are less well understood, and in part a reflection of the general attention and fashion in science at this time. I justify it by the thought that eventually the study of all periodic phenomena will profit from a thorough understanding of one. I grieve that it was not practical to include a discussion of experiments with animals since many of these experiments have been crucial, and I have tried not to omit basic concepts because of this limitation.

When writing and thinking about this book, I have imagined before me a young student who has heard that rhythms exist and has come to ask what they are and how they work. I am trying to start him in the right direction so that he will be able to do meaningful experiments of his own, understanding what has already been done, and I hope unencumbered by theoretical prejudices. People who already understand the experiments on rhythms will not find this book helpful, nor discover in it new theories or insights. I would have been glad to offer such new interpretations were I able. My hope is that the facts set forth here will make possible such a contribution by a new investigator.

Department of Biological Sciences,   B. M. SWEENEY
University of California,
Santa Barbara, California, U.S.A.
October 1968

Acknowledgements

I would like to express my appreciation to my fellow investigators in the field of rhythms, who, through discussions and criticisms, have helped me so much and whose experimental results serve to illustrate many points in this book. Especially, I should like to thank the National Science Foundation for their support of my research; Dr G. Benjamin Bouck for the beautiful frontispiece electron micrograph; Mrs Karen Chruszch for her careful typing of the manuscript; and my very patient husband, Paul Lee, for his never-failing encouragement.

For permission to reproduce in whole, or in part, certain figures, I am grateful to the following publishers:

American Society for Microbiology (Fig. 3.1); American Society of Naturalists (Fig. 5.8); American Society of Plant Physiologists (Figs 3.21, 5.1, 5.7); The City College (Fig. 6.3); Elsevier Publishing Co. (Fig. 7.4); Gustav Fisher Verlag (Fig. 8.1); Long Island Biological Association (Figs 3.15, 8.2); Macmillan (Journals) Ltd. (Fig. 3.5); National Academy of Sciences (Figs 3.6, 3.13, 3.14, 6.2); National Research Council (Fig. 5.4); Rockefeller University Press (Figs 3.18, 8.3, 8.4); Society of Protozoologists (Fig. 7.3); Springer Verlag, Berlin (Figs 3.3, 3.4, 3.20, 3.25); Springer Verlag, Vienna (Fig. 6.1); Verlag de Zeitschrift f. Naturforschung (Fig. 3.8); Wistar Institute of Anatomy and Biology (Figs 3.9, 7.1, 7.2).

Contents

First Observations: The Patterns of Plant Movement

To us self-important humans, it is rather startling to realize that animals can tell time without the aid of wrist watches. The thought that even plants can accomplish this feat is at first almost too much to swallow. However the evidence is quite clear. Plants as well as animals show behaviour patterns which repeat with a definite and accuratetime interval. Since these periodicities continue without cues from the outside world, they are essentially time-keeping devices, and serve to regulate activities with respect to the time of day, month, tide or year. The first plant periodicities were discovered very long ago. They were periodic movements of leaves.

The power to move is so firmly associated with animals that we feel a sense of surprise the first time we notice that plants too can move. Yet we are familiar with the fact that the cells of plants and animals are fundamentally alike. How much more astonished must the naturalists of long ago have been when they observed that leaves and flowers could change their position. It is no wonder that early records of plant lore contain references to the movement of plants. One of these, a record from Pliny's writings, concerns the observation that some kinds of leaves take up a different position at night from that which they occupy during the day. Carl Linnaeus, in 1751, coined the name 'plant sleep' for these movements (Fig. 1-1), and they are called sleep movements to this day.

That sleep movements represent anything other than a direct response to the light of day and the darkness of night was first suspected by De Mairan, in 1727. Appropriately enough, he was an astronomer and interested in the earth's rotation. He found by experiment that sleep movements would occur in continuous darkness, and hence the change in position was not a response to change in light conditions. His findings were confirmed by Jinn in 1759 and, in 1835, by De Candolle, who studied the leaf movements of *Mimosa pudica* and found that if the plants were placed in artificial light every night, the leaves still 'slept' and 'woke' as they would have done, had the plant been growing under the natural alternation of day and night, except that now the time

(a)

(b)

Fig. 1-1. Sleep movement of the leaves of *Albizzia julibrissin*; (a) day position, (b) night position.

from one maximum spreading of the leaflets until the next was not twenty-four hours as it would have been in nature, but shorter than this by one to one and a half hours. No particular importance was attached to this observation until much later. De Candolle also looked at plants kept in darkness, where he found the movements present but somewhat less regular than in continuous light. He found that many other leaves which showed sleep movements, *Oxalis* and *Phaseolus* for example, and even flowers which close at night continued this behaviour in the absence of changes in the light environment. In his studies of the sleep movements of *Mimosa*, De Candolle also found that it was possible to reverse the course of the movements by placing the plants in darkness during the day and artificial light at night. Thus it seemed that, although the changes in light were not the cause of the leaf movements, they still could influence them.

There now began a period of intense interest in the patterns of motion in leaves, tendrils and shoots, which touched almost every eminent plant biologist of the second half of the nineteenth century: Darwin, Dutrochet, Hofmeister, Sachs, Pfeffer. These men were concerned with how movements were executed, but also with why they existed and what caused their periodicity, whether it sprang directly from some environmental periodicity, or whether it had some other basis.

At first, there was no method for making automatic recordings of the movements of leaves, so that all observations depended on the wakefulness, as well as on the patience and tenacity of the investigator. Measurements were necessary, however, before anything more could be learned about sleep movements and their cause.

Charles Darwin was admirably suited for this kind of work, both because he was a very careful observer but also because he was a semi-invalid and suffered many sleepless nights. He kept plants of various kinds in his room, and his attention was soon caught by the movements that many of them showed. In his book 'The Power of Movement in Plants', published in 1881, he described what he had observed. He watched tendrils swing about in the air until they touched something, when they quickly coiled, and he watched many parts of plants swing slowly in circles or ellipses. But he found that the motion was too slow to observe accurately without some method of making measurements. Consequently, he devised a system for recording the excursions of plant organs. For this purpose, he attached a very light thin glass fibre to the shoot or leaf. At the free end of this fibre he fixed a small bead of sealing wax. Whenever he made a measurement, he lined up this bead with a black spot painted on a card which was placed in a convenient position behind the plant he was measuring. By this means he avoided changing the position of his eye from one measurement to another. Between his observation point and the plant he now stood a sheet of glass on which he marked the position of the bead with a wax pencil, writing the time beside it. The record was completed by joining the points in order. This method amplified the motion. One of these records is shown in FIG. 1-2. Had his points been spread out on a time axis, by moving the tracing paper an amount proportional to the time interval between readings, a rhythm would have been displayed. None the less, his measurements had the great advantage of providing an accurate record of what was happening.

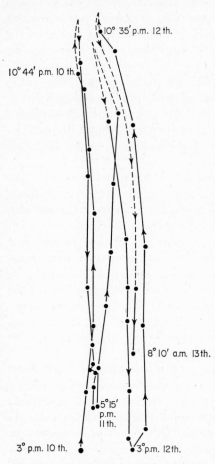

10° 35' p.m. 12 th.

10° 44' p.m. 10 th.

8° 10' a.m. 13th.

5°15' p.m. 11th.

3° p.m. 10 th.

3° p.m. 12th.

FIG. 1-2. Movement of a leaf in *Nicotiana tabacum* traced by Charles Darwin on a vertically-placed glass. (C. Darwin and F. Darwin, 1881. 'The Power of Movement in Plants.')

Darwin was primarily preoccupied with circumnutation, the waving of shoots in an ellipse. He considered sleep movements to be a specialized form of circumnutation, which he called 'nyctotropism' or night-turning. Naturally he was interested in what kind of selective advantage such movements might confer. He thought that since the upper surfaces of leaves are almost always pressed together at night in the course of these movements, whether they take place by upward or downward motion of the leaf, the night position might protect the leaves from excessive loss of heat by radiation to the cold night sky. He pinned the leaves of several kinds of plants, including *Trifolium pratense* and *Oxalis purpurea*, in the day position against a piece of cork and exposed these plants for various times outside at night. He noted that the leaves so pinned showed more damage than those of the same plant which were permitted to 'sleep'. Open leaves were sometimes more covered with dew, and he cites this as evidence that they were colder than the folded ones. He does not explain what advantage sleep movements confer on the leaves of tropical plants, although he cites as examples several tropical species, and must have been aware from his travels how prevalent sleep movements are in warm climates where leguminous trees are so predominant. He observed that the cotyledons of a wide variety of plants show sleep movements and lists these species, as well as the species which show nyctotrophic movement as adults. These lists attest to the thoroughness of his investigations, and could be useful to us today in selecting experimental objects. However, Darwin was fundamentally a very careful observer, rather than an experimentalist. He confirmed the widespread occurrence of sleep movements rather than adding new data toward interpreting their cause.

Julius Sachs, on the other hand, was first and foremost an experimental plant physiologist. In April, 1863, he selected a plant of *Acacia lophantha* which he assures us had 'new fine, very healthy-looking leaves'[1] and placed it inside a wooden box with a thermometer beside it. He then measured the angle the leaflets made with the petiole every hour during the day for four days, recording the temperature with each measurement. The plant thus kept in darkness continued to show sleep movements for about 48 hours, but after this they could no longer be measured. There was no correlation between the movements of the leaves and the temperature variation inside the box. Thus the possibility was eliminated that in darkness, temperature cycles were causing the sleep movements.

In 1873, Pfeffer published his observations on the sleep movements in the same plant studied by Sachs, *Acacia lophantha*. In these experiments, he wished to measure the behaviour of the leaves in a constant environment, but one where the plant did not become progressively more starved as it would in long-continued darkness. Consequently, he attempted to keep his plant in continuous light, which he approximated by lighting it at night with artificial light in the same manner as De Candolle had done, but with much brighter light concentrated on the leaves with mirrors (FIG. 1-3). The heat from these intense lamps was reduced by two water filters between the lights and the plant, but temperature changes over the course of 24 hours were by no means eliminated. During the day, the plant was lighted by natural light, thus the intensity was not constant. Pfeffer found, as had Sachs, that the sleep movements persisted for about 48 hours under this experimental arrangement, but gradually damped out after this. He had previously

[1] Sachs, J., Lectures on the Physiology of Plants. Trans. H. Marshal Ward, 1887. P. 636, lines 12–13.

found that the opening and closing of flowers such as Crocus and Tulip was determined by a rise or fall in temperature, and that when the temperature was suddenly raised, the flowers opened very widely, then closed to an intermediate position, sometimes showing several oscillations before coming to an equilibrium position. Perhaps by analogy with this overshooting in the opening of flowers, he at first considered

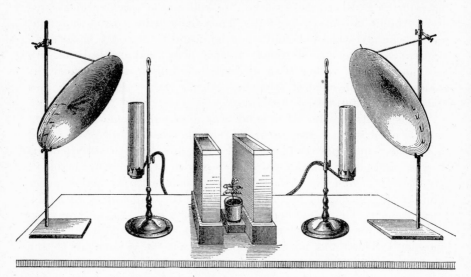

FIG. 1-3. Pfeffer's arrangement for lighting plants continuously. (Pfeffer, 1875. Die periodischen Bewegungen der Blattorgane.)

that sleep movements persisting in continuous light or darkness represented 'after effects' caused by the previous changes in light in the environment.

In 1875, Pfeffer's book 'Die Periodischen Bewegungen der Blattorgane' was published, summarizing all his early work on this phenomenon, as well as giving an extensive review of the work of other investigators of the time. In it, he elaborates his theory that sleep movements are essentially after-effects of the light-dark cycle of the environment, paying much attention to the dying away of the movements in constant light or darkness. One of his experiments showing the leaf movement in *Acacia* in darkness is shown in FIG. 1-4. He notes that there are differences from plant to plant in the length of time that the leaf movements can be observed. Furthermore, the dying away is not due to loss of 'irritability' since he found that plants which had lost rhythmicity in continuous light immediately assumed the night position when darkened. He was also able to induce a new periodicity in

Sigesbeckia orientalis which had lost its original rhythm after being in continuous light for 5 days. This new rhythm was brought about by darkening the plant from 8 a.m. to 4 p.m. and hence was reversed with regard to natural night and day.

For some interesting reason, Pfeffer himself apparently never performed the experiment crucial to a test of his theory that after-effects

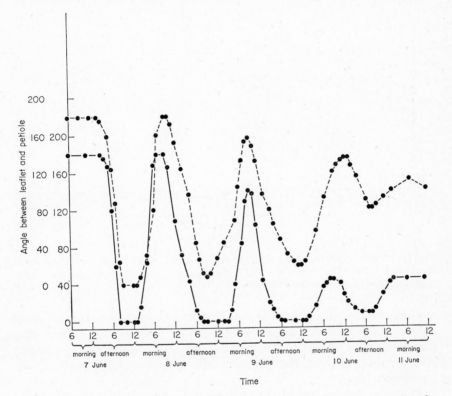

FIG. 1-4. Pfeffer's measurements of the sleep movements of an old (upper curve) and a young leaf (lower curve) of *Acacia lophantha,* in continuous darkness. (Pfeffer, 1875. Die periodische Bewegungen der Blattorgane.)

explained the persistence of rhythms in constant conditions, namely to grow a plant in an artificial light-dark cycle in which the sum of the light and dark periods was not 24 hours, then move the plants to constant conditions and observe whether the after-effect reflected the imposed light-dark cycle. This experiment was soon performed by Semon, however.

In 1905, Richard Semon published very convincing curves which showed that seedlings of *Acacia lophantha* which had been grown from germination in darkness or in light-dark cycles of 6 hours of light followed by 6 hours of darkness, or 24 hours of light followed by 24 hours of darkness, still invariably showed leaf movement which followed a 24-hour pattern rather than a 12- or a 48-hour cycle, after the plants were subsequently placed in continuous light. Figure 1-5 is one of his records from a plant on a 6-hour schedule. Semon found that plants in

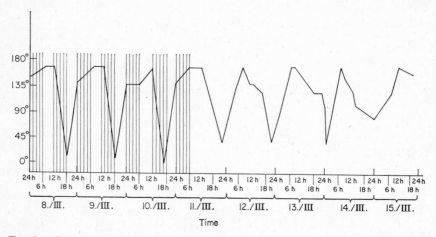

Fig. 1-5. Semon's measurements of the sleep movements of leaves of *Acacia lophantha* in LD 6 : 6 followed by continuous light, showing that after-effects of the light-dark regime are not detectable. (Semon, 1905.)

continuous darkness did not repeat their leaf movements exactly every 24 hours but required somewhat longer for a full cycle so that the maximum extension of the leaflets did not always occur at any certain time of day. He concluded that these diurnal movements had an hereditary basis.

Pfeffer at first criticized the experiments of Semon, saying that perhaps some daylight was leaking into his experimental apparatus. However, for his own later experiments, Pfeffer devised an ingenious method for recording leaf movements automatically. Figure 1-6 shows the arrangement of his apparatus. Note that such an arrangement writes the record upside down, the pen dropping when the leaf rises. His new experiments convinced him that Semon was correct and that the leaf movements were indeed inherited rather than being mere after-effects.

Leaf sleep-movements were not the only periodic responses of plants known to botanists of the later part of the nineteenth century. Sachs performed some amazingly modern experiments which demonstrated the existence of rhythms in growth and in the exudation of fluid from the cut shoots of plants. Since growth showed a maximum every night, he at first considered that light was the determining factor, since bright light was known to retard growth. However, like the sleep movements, the periodicity of growth also occurred in continuous darkness. He at first doubted that he had excluded all light but was reassured when Baranetzky repeated his growth experiments in more elaborate form in

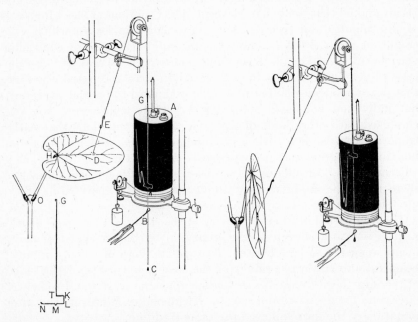

FIG. 1-6. Device for recording leaf movements automatically, similar to that devised by Pfeffer and used by Brouwer and Kleinhoonte. On the left the leaf is shown in the day position, on the right in the night position. (Brouwer, 1926.)

1879. Although Baranetzky had shown that diurnal variations in growth occurred in etiolated shoots of *Brassica rapa* grown from tubers in darkness, he attributed the diurnal growth patterns to after-effects of a periodic environment as had Pfeffer. Sachs could not ascribe to this view, which he regarded to be 'much as if a pendulum commenced to swing spontaneously after it had been standing still for some time'[1]. He concluded that the periodicity arose from purely internal causes which were modified by the alternation of day and night to produce a daily period. In this, his opinion was strengthened by an experiment on the rate of exudation from the cut stump of a sunflower growing in his

[1] Sachs, J., Lectures, p. 560.

garden. He noticed that there were regular fluctuations in the rate of exudation from a maximum of 14 ml per hour at 8–9 a.m. to a minimum of 2·5 ml per hour during the night. This periodicity did not depend on temperature variations nor on the moisture content of the soil. He stated quite frankly in his lectures on plant physiology (p. 273) that 'the cause of the daily periods is simply unknown'.

Biological cycles which required either less than or more than 24 hours were only mentioned here and there in the older literature. Darwin noted that some plants execute jerky motions at short intervals of 4–6 minutes and recorded such movements in leaves which were closing at night. Pfeffer alludes to the rapid pulsation of contractile vacuoles in *Ulothrix* (12–15 sec) and *Gonium* (26–60 sec). He also discusses the annual changes in bud growth in trees and noted especially that pear trees transplanted to Java were known to become evergreen, but individual buds still showed rest periods which were not synchronized with other buds of the same tree. He mentions that native tropical trees also lose their leaves and go through a short rest period but this rest period does not necessarily occur at the same time even in adjacent trees. Considering these observations, Pfeffer speculates that there might be an hereditary annual periodicity apart from effects brought about directly by the changing seasons.

The first period of intense activity in the study of plant periodicities really came to a close with the admission by Pfeffer that leaf movements appeared to be inherited. Two papers, both of which contained beautiful automatic recordings of the sleep movements of the leaves of the large bean *Canavalia ensiformis* appeared in Holland somewhat later, one by Brouwer and one by Kleinhoonte. However, the interpretation of the data followed the same tradition.

Let us summarize the accomplishments of the plant biologists of the nineteenth century. Periodic leaf movements, and to some extent, periodic growth and exudation, were measured under natural conditions, and in both continuous darkness and continuous light, with some attention given to the ambient temperature. The results of these experiments showed that cycles, about 24 hours long, continued for a time without alternation of light and darkness, then disappeared. This was finally unanimously interpreted as an indication of the hereditary nature of the periodicities. Some experiments with plants in light-dark cycles longer or shorter than 24 hours had been performed, and it was known that these cycles are not learned. Although they may be followed while in force, after such cycles are stopped the plants at once return to their diurnal behaviour in either constant light or continuous darkness. That light did have an effect on periodicity was known from experiments where day and night were reversed artificially, but the details of light effects had not been examined. The lack of recording methods during most of this time perhaps accounts for the failure to emphasize the observation that the length of the period of rhythms in constant conditions was often not exactly 24 hours. No one even hinted that rhythms might be indications that plants can distinguish one time from another, possess a chronometer if you will.

Following the last papers of this period, there occurred a lull in research on rhythms which lasted almost without interruption for twenty years or so. Paradoxically, this quiescent period coincided with a leap forward in the technology necessary for the control of the environment. Truly constant light and temperature conditions could now be achieved in the laboratory, and physiologists embraced this chance to control not only the conditions during the course of an experiment but also during the growth of their experimental material. Constant light was the order of the day. They strove to grow plants which were extremely uniform without variation from plant to plant or *from time to time*, and so they regarded variation in the time domain as a result of failure in their experimental technique. Such an attitude does not welcome the discovery of rhythmicity. Indeed, rhythmicity was not usually present since the favourite environment was one of bright constant light, a condition unfavourable for rhythms as we shall see.

Harbingers of a renewed interest in periodicity were the papers of Bünning, the first of which was published in 1930. His work with the leaf movements of *Phaseolus* revived interest in the diurnal leaf movements, but far more important was his realization that these leaf movements might be manifestations of a fundamental ability to tell time. In later papers, he invoked this time measurement in his theory to explain how some plants can detect long or short days, and respond to their presence by taking on the flowering condition, the phenomenon of photoperiodism discovered in 1920 by Garner and Allard. With this insight of Bünnings's, the timekeeping aspect of plant periodicities came into prominence, and the recent wave of research on rhythms was set in motion.

References

Baranetzsky, J. (1879). Die tägliche Periodizität im Längenwachstum der Stengel. *Zap. imp. Akad. Nauk*, 7th ser., **27**, 1–91.

Brouwer, G. (1926). De periodieke Bewegingen van de primaire Bladeren bij *Canavalia ensiformis*. 110 pp. J. H. Paris, Amsterdam.

Bünning, E., and Stern K. (1930). Über die tagesperiodischen Bewegungen der

Primarblatter von *Phaseolus multiflorus*. II. Die Bewegungen bei Thermo-kanstanz. *Ber. dt. bot. Ges.* **48**, 227–252.

De Candolle, A. P. (1832). *Physiologie végétale.* Paris.

Darwin, C. (1876). The movements and habits of climbing plants. D. Appleton and Co., New York.

Darwin, C., and Darwin F. (1881). The power of movement in plants. D. Appleton and Co., New York.

Dutrochet, H. (1837). Mémoires pour servir à l'histoire des végétaux et des animaux. Brussels.

Hofmeister, W. (1867). Die Lehre von der Pflanzenzelle. Wilhelm Engelmann, Leipzig.

Kleinhoonte, A. (1929). Über die durch das Licht regulierten autonomen Bewe-gungen der Canavalia-Blätter. *Archs. néerl. Sci.*, ser. 3b, **5**, 1–110.

Linnaeus, C. (1751). Philosophia botanica. 274 pp. Stockholm.

Linnaeus, C. (1755). Somnus plantarum. Amoenitat. *Academicae* **4**, 333.

De Mairan, M. (1729). Observation botanique. p. 35. Histoire de l'Académie Royale des Sciences, Paris.

Pfeffer, W. (1875). Die periodischen Bewègungen der Blattorgane. 176 pp. Wilhelm Engelmann, Leipzig.

Pfeffer, W. (1911). Der Einfluss von mechanischer Hemmung und von Belastung auf die Schlafbewegungen. *Abh. sächs. Akad. Wiss.*, Math-Phys. Kl. **32**, 161–295.

Pfeffer, W. (1915). Beiträge zur Kenntnis der Entstehung der Schlafbewegungen. *Abh. sächs. Akad. Wiss.*, Math-Phys. Kl. **34**, 1–154.

Plinius. *Hist. Natural. Lib.*, **18**, chap. 35.

Sachs, J. (1857). Über das Bewegunsorgan und die periodischen Bewegungen der Blätter von Phaseolus und Oxalis. *Bot. Zeit.* **15**, 809–815.

Sachs, J. (1863). Die vorübergehenden Starre-Zustände periodisch beweglicher und reizbaren Pflanzenorgane. II. Die vorübergehende Dunkelstarre. *Flora* **30**, 465–472.

Sachs, J. (1887). Lectures on the physiology of plants. Trans. H. M. Ward. Clarendon Press, Oxford.

Semon, R. (1905). Über die Erblichkeit der Tagesperiode. *Biol. Central.* **25**, 241–252.

Semon, R. (1908). Hat der Rhythmus der Tageszeiten bei Pflanzen erbliche Eindrücke hinterlassen. *Biol. Central.* **28**, 225–243.

Zinn, J. G. (1759). Von dem Schlafe der Pflanzen. *Hamburgisches Magazin* **22**, 49–50.

CHAPTER 2

A Short Dictionary for Students of Rhythms

In any field there are certain ideas which recur very frequently. Sometimes the expression of these ideas in every-day language requires many words and awkward explanatory phrases. For this reason, a special vocabulary soon develops, a convenient shorthand which may be obscure to the uninitiated. In the field of rhythms this is particularly true, since the recurrent ideas are somewhat more than ordinarily difficult to express in common parlance. Many observations are necessary to demonstrate the presence of a rhythm and one often wishes to describe all these data together in the form of a graph, with time as the abscissa. Then, too, it is often necessary to refer to the light conditions where a light period of say twelve hours alternates with a dark period of the same or of a different length and is followed by constant light. Consequently, shorthand expressions have been developed to describe both the conditions of rhythm experiments and their results, thought of graphically. A short explanation of these terms seems worthwhile including here.

Perhaps the first thing to define, so that we shall be quite clear what we are talking about, is the term *biological rhythm*. A *rhythm* is a regular fluctuation in something, e.g. in the position of a leaf, like the rhythms discussed in Chapter 1, or in the rate of a physiological process, photosynthesis for example, or in shape, in the production of spores, in colour, in activity. The important thing is that a *repeating* pattern can be discerned clearly. However, every repeating pattern is not a rhythm. A rhythm is self-sustaining and it is important to understand what this implies. Perhaps an example will make this clear. Let us say that we observe that the leaves of a tree in the garden are folded together every night but spread wide again every morning. There is a repeating and predictable pattern in this behaviour, but is it a biological rhythm? We cannot tell, since the enormous difference between night and day which is itself a repeating pattern may be *causing* the differences which we see in the leaves. They may open in light and close in darkness. If this is the case, the change in leaf position is not a biological rhythm but simply a response to light, and the rhythmic pattern of the leaf position is not self-sustaining, and would disappear at once were the environmental changes removed. Just this experiment must be done to detect the presence of a true rhythm. It must be shown to be self-sustaining, at least for a time, in constant conditions of light and temperature.

Before we consider the terms which are used to describe rhythmic phenomena, let us pause to think about a repeating pattern with which we are familiar, the turning of a wheel. The name *cycle* comes from this analogy where a circle is changing its position in time, yet any

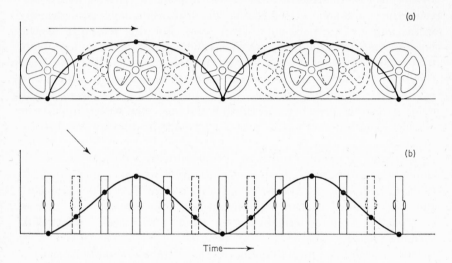

FIG. 2-1. Derivation of cycles from the turning of a wheel.

point on the rim is always returning to contact the ground. There are several ways in which we can picture what is happening to points on this wheel, as it turns. From a vantage point to the side of the direction in which the wheel is rolling, a point on the rim will appear to follow a scalloped course (FIG. 2-1a), while looking at the wheel from in front, the same point will describe a sine curve (FIG. 2-1b). These curves are cyclic and come, as we see, from a circle in motion in time. Their pattern repeats over and over. It is an *oscillation*. The fundamentally oscillatory character of rhythms makes it appropriate to use for them the same terms used in describing physical oscillations like that of the wheel, the pendulum, or the electronic oscillators. The time required for an oscillation to make a complete cycle and return again to the starting position is the *period*. In the wheel analogy, it is the time for one rotation of the wheel. In the rhythm in leaf movement, where leaves are spread open maximally every day at noon, the period is the time from noon to noon or 24 hours. Note, though, that the period is the same no matter what point in the cycle is used to measure it.

As in physical oscillations, the inverse of the period is the *frequency*, the number of cycles which take place in a given time. In the discussion of biological rhythms, frequency is used less often than period, however. Just as in physical oscillations, the frequency of rhythms may be forced to match exactly that of some external oscillation, the alternation of light and darkness for example. When this occurs, the rhythm is said to be *entrained* by this external oscillation. The signal responsible for entrainment has been called the 'zeitgeber' or 'time-giver'. Entrainment, of course, can only take place while the external oscillation is actually present. When a rhythm is *not* entrained, it is said to be *free-running*, and now shows its *natural* period.

Under certain special conditions, oscillations can be entrained to show periods which are a multiple of the entraining cycle. This phenomenon is called *frequency demultiplication* since the entrained period which the rhythm shows is longer than that of the external cycle which is entraining it.

Amplitude is the term which is used to describe the intensity of an oscillation. A rhythm with a large amplitude is one in which there is a big difference between the maximum and the minimum in the quantity which is showing the rhythmic variations.

In discussing a process that is changing in a rhythmic manner, not only do we need a term for the time that it takes to make one complete cycle and return to the starting point (the period), but it is necessary also to describe just where in this cycle the rhythm is at any certain time. Is the process at its maximum, minimum, ascending or descending in rate? For the stage of the cycle the term *phase* is used. This term is often employed to describe where the cycle is relative to some other time basis, say sidereal or solar time. Two rhythms may be said to be out of phase when their maxima do not coincide. It is also possible, as we shall see, to change the phase of rhythms in various ways. This means simply changing the solar time at which some phase occurs, moving the maximum from noon to midnight for instance. Since oscillations are really projections of circular motion as we saw in the analogy with a rotating wheel, the phase may be described in degrees, and so the phase of one rhythm relative to another may be measured in degrees, as '90° out of phase'. Changes in phase are also measured in this way, a change which results in a maximum where a minimum formerly occurred is then a 180° phase shift or a half of a circle. Thus the phase may be changed by moving the whole cycle forward or backward in time, *without changing the period*. Changing the phase of a rhythm is also called 'resetting' the rhythm by analogy with moving the hands of a clock. When alterations in phase are produced, it sometimes happens that the complete change does not occur in a single cycle, but may take several cycles before the final new phase is established. During this transition, the apparent period will be either shorter or longer than usual and will be unstable, perhaps changing with each cycle. Such intermediate cycles are known as *transients*.

When the new maximum occurs earlier than the old, the phase is said to be changed in the positive direction, while delay in the maximum is regarded as negative. Curves in which the change in phase is plotted as a function of when in the cycle a single stimulus was applied are called *response curves*, and values for the amount of phase shift are either positive or negative according to this convention.

In physical systems, oscillations usually take on a wave form. Although some rhythms also appear to follow a wave pattern, this is by no means universal. Whatever the apparent shape of the biological oscillation, however, the terms period, phase and entrainment are used.

Biological systems which oscillate may be classified in various ways. Biological rhythms in the true sense of the word are only those oscillations which can be shown to continue in the absence of periodic changes in the environment, such as changes in light and temperature, i.e. self-sustaining oscillations. Such rhythms may be called *endogenous* since they apparently arise from inside the organism. Aschoff, Klotter, and Wever refer to them as *active systems* since they are *self-sustained oscillations* capable of deriving the energy necessary for their maintenance from constant sources of energy. This is in contrast to *passive systems* which appear rhythmic by virtue of their ability to respond to periodic variables in the environment, and cannot derive energy for their continuation from any constant source. Such periodicities are also called *exogenous*, since their origin is outside the cell. They are not true rhythms, since they are not self-sustaining. Since many true rhythms gradually die out or *damp out* in constant conditions, it is not always easy to distinguish a true rhythm from one of these exogenous periodicities, especially since we are not always sure that the environment does not contain some periodic variable to which the organism can respond, perhaps one of which we the experimenters are unaware.

Biological rhythms may also be classified according to the length of the period they show, as *diurnal* or *circadian* when the period is about 24 hours, *tidal* when the period approaches 12·8 hours, *lunar* if 28 days and *semi lunar* if 14–15 days, *annual* if one year.

The most frequently used of the shorthand expressions that have been devised are those which describe the conditions of light or darkness used in experiments on rhythms. Alternating light and darkness is represented by L (light), D (darkness), and the number of hours of each is written directly afterward, light duration first, then darkness, both in hours. A light-dark schedule in which 12 hours of light is followed by 12 hours of darkness and this routine is repeated over and over is written as LD 12 : 12 for example, or if the light period is 16 hours and the dark 8, as LD 16 : 8. The sum of the hours of light and darkness need not be 24, of course, as for example LD 8 : 8. The light intensity in lux may be written in brackets after the hours of light and darkness, as LD 24 : 24 (5000). In this notation, LL 100 designates continuous light of 100 lux intensity, and DD, continuous darkness. Sometimes other symbols are also used, especially by Aschoff and his co-workers (1965). In his notation, the Greek letter tau (τ) stands for the period, and phi (ϕ), the phase. Phase shifts or changes are written as $\Delta\phi$. Small letters are used when talking about the biological oscillations or rhythms while capital letters denote the period or phase of environmental oscillations. In work with rhythms in animals, where activity is the rhythmic variable, alpha (α) is used for the activity time or the part of the cycle when the animal is active, while rho (ρ) denotes rest time. The alpha : rho ratio is sometimes of interest. By analogy alpha may also sometimes be used to denote the time in the cycle when a reaction is fastest or a process other than activity is taking place at the highest rate, and rho the reverse, but in plant rhythms this concept has less meaning.

These are the principal terms which we will need to describe rhythms. With tools thus properly sharpened, we are now ready to tackle the subject of what is known of these rhythms in biological systems, and what is still to be discovered.

REFERENCES

Aschoff, J., Klotter, K., and Wever, R. (1965). Circadian vocabulary. Pp. x–xix. *In* Circadian Clocks (ed. J. Aschoff), North Holland Publishing Co., Amsterdam.

CHAPTER 3

Rhythms that Match Environmental Periodicities: Day and Night

From its beginning, the earth has been in rotation and the alternation of day with night, the result of this rotation, has dominated the environment of our planet. All living things have evolved under this inviolate periodicity, and it is not surprising that they are indelibly imprinted with its frequency. In fact, rhythms with a period close to 24 hours come to light everywhere in the biological domain. So common are they that some scientists believe them to be present in every living cell. Although diurnal rhythms have not been found in every organism where they have been sought, the possibility remains that we have not examined the right processes for their presence.

Rhythms in which the period is of the order of 24 hours were at first called diurnal rhythms. Some investigators felt however that this might have been a poor choice, since a confusion arises because 'diurnal' means in the daytime, and is often used as the opposite of nocturnal. To avoid ambiguity, Halberg suggested that these rhythms be called 'circadian', from the latin *'circa'*, about, and *'diem'*, day. This term is appropriate since it stresses one of the most important properties of such rhythms, that the period is often not exactly 24 hours. It has come into general usage in spite of a slight awkwardness arising from the similarity between circadian and cicada, especially when these words are pronounced by Bostonians such as myself.

Many circadian rhythms have now been investigated, and the list is daily growing. It includes organisms at many levels of complexity: single-celled flagellates like *Euglena* and *Gonyaulax*, *Paramecium*, filamentous algae like *Oedogonium*, moulds, Crustacea, many insects, higher plants, birds, mammals including man. Notable exceptions to this list are the prokaryotic cells. Circadian rhythms in bacteria and blue-green algae are unknown with one possible exception. This exception involves the advance of a culture of *E. coli* down a long tube filled with medium, investigated by Rogers and Greenbank (1930). Pronounced maxima which vary somewhat in spacing but average 20·9 hours apart are clear

18

in the data for this advance (FIG. 3-1). The question whether or not prokaryotic cells can show circadian rhythms is interesting when the mechanism of these rhythms is under consideration.

In plants, a relatively small number of circadian rhythms have been investigated in enough detail so that their characteristics are well known. Since almost all that we understand about rhythms in plants is

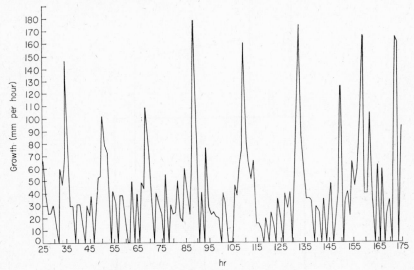

FIG. 3-1. The rate of advance of a culture of *Escherichia coli* down a race tube. Note the distinct peaks which occur about 20 hours apart. (Rogers and Greenbank, 1930.)

based on these systems, it seems worth while to describe in some detail what they are and how they are measured.

A very large part of the work of Bünning and others in his laboratory has been done using the rhythm in leaf movement in the common bean, *Phaseolus multiflorus*. Seedlings are used, and the change in the angle which the midrib of the primary leaf makes with its petiole is measured automatically. This is accomplished by essentially the same method as that used by Pfeffer and by Brouwer and Kleinhoonte. Other parts of the plant beside the leaf blade to be measured are secured in a fixed position. A light thread is attached to the leaf blade, and runs to the short arm of a lever or over a pulley (FIG. 3-2). If the lever is used, the long arm carries at its tip a pen which writes on a kymograph drum. This drum turns at a very slow rate, requiring a week or so to make one revolution. If a pulley is used (FIG. 3-6), the free end of the thread carries a weight and a pen is attached between pulley and weight. The lever arrangement has the advantage that the movement of the blade is magnified, but its motion is in an arc. This is avoided in the pulley arrangement, which does not magnify the motion however. Both methods of recording reverse the direction of the leaf movement, so that a maximum on the record represents the lowest or night position

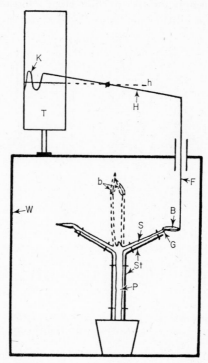

Fig. 3-2. Bünning's apparatus for recording the sleep movements of *Phaseolus multiflorus*. Movement of all parts of the plant except the leaf blade are prevented by restraints. A lever (H) is attached to the blade by a thread (F), and carries a pen at the other end which writes on a rotating kymograph. The plant is enclosed in a box in which temperature and light can be controlled. (Bünning, 1931.)

of the blade. Hoshizaki and Hamner (1964) have recorded the bean leaf movement rhythm using time-lapse photography.

The site of the changes which alter the position of the leaf is the pulvinus, a specialized tissue at the leaf-petiole junction, the cells of which have a changing turgor pressure with time in the circadian cycle. Fig. 3-3 is an example of Bünning's recordings of the leaf movement of *Phaseolus multiflorus* in constant light. Beans which have been grown since germination in constant white light do not show any leaf movement rhythm until some change in the environment sets it in motion. All that is required is a single 9–10-hour exposure to darkness. Once set in motion, this rhythm will persist in constant light or darkness for at least 6–8 days, with a period of about 28 hours. Hoshizaki and Hamner report the persistence of a similar leaf sleep-movement rhythm for at least 28 days in *Phaseolus vulgaris* var. Pinto in LL as

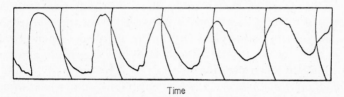

Time

FIG. 3-3. A record of the leaf movements of the primary leaf of *Phaseolus multiflorus* in continuous light and 21–22° C. Vertical lines have been drawn every 24 hours. (Bünning and Tazawa, 1957. *Planta* **50**, 107–121).

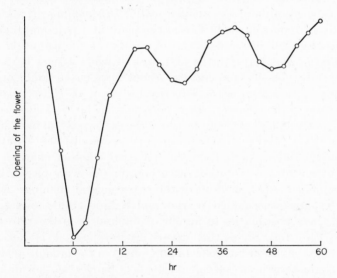

FIG. 3-4. Measurements of the opening and closing of flowers of *Kalanchoë blossfeldiana* in continuous darkness. (Bünning and Zimmer, 1962. *Planta*, **59**, 1–14.)

bright as 4500 to 7000 lux; the period in this case is also longer than 24 hours (Ottmans, 1960; Karvé *et al.* 1961).

Bünsow (1960), Engelmann (1960) and Zimmer (1962), of Bünning's laboratory, have examined another movement rhythm, that of the petals of *Kalanchoë blossfeldiana*. *Kalanchoë* is interesting because it is a short day plant and details concerning its photoperiodic responses are known. The flowers are small, consisting of four petals joined to form a tubular corolla. At night, the petals move up and in, closing the flower, while during the day, they move down and away from each other to open the flower. This movement continues in flowers that have been detached from the plant and mounted separately in holders provided with water. Recordings can be made by time-lapse photography, the camera being placed so that it looks down on the flowers from above. The degree of opening can then be measured from the pictures as the distance across the petals. The rhythm in petal movement continues for about three cycles in DD (FIG. 3-4). When the rhythm damps, the flowers are in the open position.

Bryophyllum fedtschenkoi, a succulent plant somewhat similar to *Kalanchoë* in appearance has been studied by Wilkins in England. Like other succulents, *Bryophyllum* has the ability to fix larger than the usual amounts of carbon dioxide in darkness. Using an infra-red ana-lyser, Wilkins has been able to measure automatically the CO_2 changes in the gas space around an enclosed leaf of *Bryophyllum*. His standard procedure is to detach the leaf from the plant at 4 p.m. and place it in the analysing chamber in darkness. A rhythm in the evolution of CO_2 is initiated by this transfer, the first peak of which occurs 17–25 hours later depending on the temperature. About three peaks can be observed before the rhythm disappears. This rhythm is not a result of changes in the aperture of the stomata, since Wilkins was able to demonstrate its presence in mesophyll tissue from which the epidermis had been stripped (Wilkins, 1959), and in tissue culture of mesophyll cells (Wilkins and Holowinsky, 1965). At first, it was not clear whether changes in the rate of utilization or evolution of carbon dioxide were responsible for the rhythm in the CO_2 content of the experimental chamber. Analysis with $C^{14}O_2$ however enabled Warren and Wilkins (1961) to distinguish between these possibilities in favour of differences in the rate of CO_2 fixation (FIG. 3-5). Since this occurs in darkness, it is not photo-synthetic CO_2 fixation but reflects differences in the organic acid meta-bolism characteristic of succulent plants. Changes in the underlying enzyme systems have been investigated by Warren.

The phototactic rhythm in *Euglena gracilis* has attracted considerable attention since its discovery in 1948 by Pohl. Bruce and Pittendrigh (1956), devised a method for measuring this rhythm automatically, utilizing a test beam which passes through the suspension of *Euglena* held in a Carrell flask and falls on a detecting photocell. If the *Euglena* is responding phototactically, the cells will move into the beam during the 20 minutes that the test light is on and thus reduce the light reaching the photocell. The record of the output of the photocell will thus show changes proportional to the number of cells collecting in the

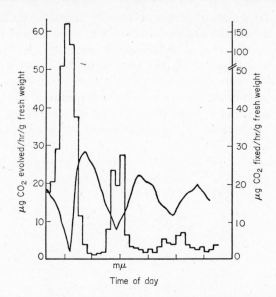

FIG. 3-5. Automatic recording of the CO_2 given off by a leaf of *Bryophyllum fedtschenkoi* in continuous darkness at 26° C (continuous curve) and the measurements of fixation of $^{14}CO_2$ by similar leaves under the same conditions (Bar diagram). (Warren and Wilkins, 1961.)

test beam. In the recordings of Bruce and Pittendrigh, the pen moves *up* when the output of the photocell is reduced and the cells are reactive. The recorder is turned off for the two hours or so between exposures to the test light while the cells are in darkness. The consequence of this method of recording is a series of rising lines, the magnitude of the rise being proportional to the phototactic response. The envelope enclosing all these lines displays the rhythm (FIG. 3-6). Recordings from successive 2-day intervals are customarily printed one below the other, overlapping the records by one day. In this method of display, the period can be detected readily from the slope of the line joining successive maxima. Accumulation in the test beam is greatest in the middle of the day phase. The strain of *Euglena gracilis* which Bruce uses is not rhythmic when it is grown on media containing organic substrates; on inorganic media, it shows cycles for about five days at the most in DD (with the exception of the test light). Brinkmann has recently measured this rhythm in *Euglena* in a slightly different way which enables him to distinguish changes in phototaxis from changes in motility, not possible in the method of Bruce and Pittendrigh. In his experimental arrangement which resembles that of the Princeton group in principle, the test beam passes through the cell suspension at a point above the bottom of the container, rather than passing vertically through the cell suspension. He finds that while phototaxis does vary, the largest part of the cyclic measurements are due to changes in motility, so that at night most of the cells are resting on the bottom of the vessel while during the day they are evenly distributed after a time in darkness.

There is one disadvantage inherent in a phototactic rhythm if it is to be used as a model system for rhythms in general: it cannot be measured without at least a brief exposure to light. By varying the

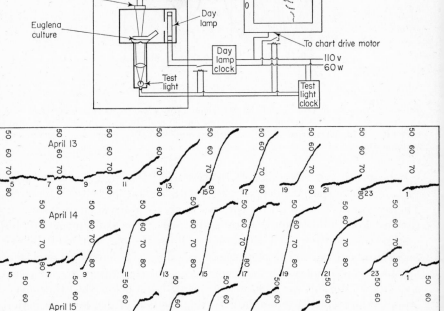

FIG. 3-6. Recordings of the phototactic accumulation of *Euglena gracilis* and a diagram of the apparatus used to make the recording. The test light was turned on for 30 minutes every 2 hours. The 'day' lamp provided LD, 12 : $\overline{12}$ as shown by light and dark bars on the abscissa of the recording. The temperature was 20° C. (Bruce and Pittendrigh, 1956.)

timing of the test light exposure, Bruce and Pittendrigh satisfied themselves that the test light did not usually reset the phase of the rhythm. However certain regularly spaced illuminations with the test light sometimes resulted in a period of exactly 24 hours in darkened cultures which under other test light schedules showed periods less than 24 hours. Bruce and Pittendrigh interpreted this finding as an example of frequency demultiplication, a phenomenon known in physical oscillations where the period can be a multiple of the interval between periodic stimuli. To avoid this complication, regular timing of the test light to exactly every 2 hours was avoided. With Brinkmann's strain of *Euglena* and experimental arrangement, a test light is also necessary but the illumination can be very brief, only long enough to secure a measurement of the dispersion of the cell suspension. Furthermore, he reports that his strain of *Euglena* also is rhythmic in media containing organic substrates, and hence able to survive long times in darkness. The rhythm under these conditions persists much longer than 5 cycles in this strain.

Könitz (1965) has examined electron micrographs of *Euglena* fixed in the middle of the light period or the dark period of LD 12 : 12 and reports that at midday pyrenoids are present and mitochondrial cristae are unoriented while at midnight pyrenoids are not to be found while the mitochondria show parallel cristae. However, these changes are probably associated with light and darkness directly and need not be manifestations of circadian rhythmicity as Könitz himself points out.

The sporulation rhythm of the green alga *Oedogonium cardiacum* was intensively studied by Bühnemann. Sporulation of this alga must be induced by a change of medium, but this change does not affect the phase of the rhythm which depends on the light-dark cycle previously in force, the maximum number of spores being released in the middle of the day phase. This rhythm continues for about 3 cycles in LL, and is measured by counting the number of germinating spores which settle on coverslips replaced every three hours. Ruddat (1961), has extended our information of the *Oedogonium* rhythm using Bühnemann's techniques (FIG. 3-7). No method of automatic recording has been developed for this rhythm.

Recently, a rhythm in the rate of photosynthesis in the large single-celled green alga *Acetabularia* has been examined by Sweeney *et al.* (1961, 7), by Schweiger and his collaborators, by Richter (1963), and by Vanden Driessche (1966). Either oxygen evolution or the incorporation of $C^{14}O_2$ may be used. This rhythm persists in constant light for very long times (FIG. 3-8), Schweiger *et al.* (1964) found it still present

B

after 40 days. Furthermore, enucleated fragments are as rhythmic as whole plants. In *Acetabularia mediterranea*, Vanden Driessche detects accompanying alteration in the shape of the chloroplasts, which she

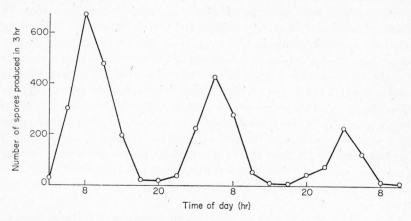

FIG. 3-7. Measurements of the sporulation rhythm in *Oedogonium cardiacum* in continuous light at 25° C. (Ruddat, 1961.)

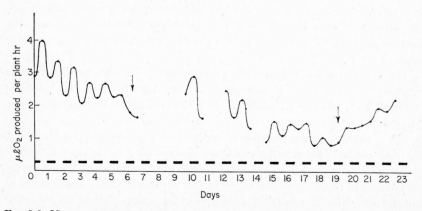

FIG. 3-8. Measurements of the rhythm in oxygen production in *Acetabularia mediterranea* with a nucleus in constant light (2500 lux). (Schweiger *et al.* 1964.)

measures as a change in the ratio of their length to their width. Swelling and shrinking has been reported to accompany photosynthetic phosphorylation, so such alternations in shape might be expected to accompany differences in the rate of photosynthesis. However, they cannot be detected in all *Acetabularia* species, perhaps because there is too great variability among the chloroplasts. The rhythm in photosynthesis is measurable both at low and at high light intensities and so does not apparently involve changes in photosynthetic capacity only, as Hellebust *et al.* pointed out.

The best known of the rhythms in algae are probably those in the dinoflagellate *Gonyaulax polyedra*. *Gonyaulax* is a unicellular flagellated creature that contains functional chlorophyll, and thus may be included among the plants. Unlike any higher plant however, it can produce light by bioluminescence, for which it can synthesize its own dinoflagellate-specific luciferin and luciferase. The cells give a flash of light when the sea water in which they are swimming is given a sharp shake or stirred vigorously. When the luminescence is measured while stirring the culture, using a photomultiplier photometer and eliciting light emission with a stream of air of sufficient flow rate for maximum stimulation, the amount of light which the cells emit is not always the same, but depends upon their recent history. If they have been growing with natural illumination or in an LD cycle, the amount of light emitted will be markedly dependent on the time when measurements are made. If luminescence is elicited during the day, the amount of light will be low, and very hard stirring will be required to bring about any luminescence at all. But if the cells are stimulated at night, a great deal more light will result and the most delicate shake will produce a flash. When the amount of luminescence is plotted as a function of time of day, a graph like that of FIG. 3-9a will result. The greatest luminescence will be produced in the middle of the dark period, while towards morning, flashes will gradually become smaller and a greater stimulus will be required. Such a finding is of course not at all sufficient to demonstrate the presence of a rhythm. That a circadian rhythm is indeed present is shown by the persistence of the changes in the brightness of luminescence when the cells are kept in darkness, FIG. 3-9b. In DD, cycles continue for sometimes as long as four days, but the amplitude becomes successively smaller.

The damping out of the rhythm in luminescence in DD can be traced to the absence of photosynthesis. Short exposures to light during the day phase increased the luminescence of the following night, the effect being proportional to the amount of illumination and the action spectrum was that of photosynthesis. Apparently some component of the luminescent system requires the products of photosynthesis for its synthesis or activation. We do not know what kinds of molecules these components are, although both the luciferin and luciferase from *Gonyaulax* have been partially purified by Hastings and Bode (1962).

Since it is apparent that the ability to luminesce is gradually lost in

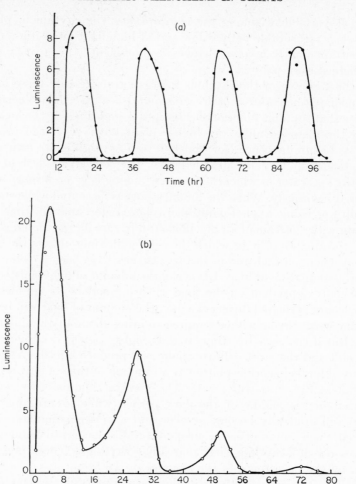

F1G. 3-9. The rhythm of stimulated luminescence in *Gonyaulax polyedra,* measured in LD 12 : 12 (a), and in DD (b). (Sweeney and Hastings, 1957. *J. Cell comp. Physiol.* **49,** 115–128.)

DD due to the absence of photosynthesis, one might expect that a somewhat different time course would be observed in continuous light. This expectation is indeed fulfilled, provided the light intensity is below about 1500 lux (Fig. 3-10). The reduction in amplitude observed in DD is no longer present, although all maxima are somewhat lower than in LD. Cycles continue for at least three weeks in LL of appropriate intensity. Recent measurements show cycles in cells kept in LL of low intensity for three months. The period can be measured quite accurately in LL and is clearly different from 24 hours at most temperatures. A long persisting circadian rhythm in luminescence is thus amply demonstrated in *Gonyaulax*.

We observe the rhythmicity of luminescence in cultures of *Gonyaulax* where of course thousands of cells are present in the 2 ml samples used for each measurement. Is the rhythmicity the result of interaction between these cells? When cultures of different phases are mixed, the luminescence of aliquots from this mixture is only the sum of its component parts. No interaction can be detected. Many other bits of evidence substantiate the view that rhythmicity is a property of the cell as an individual.

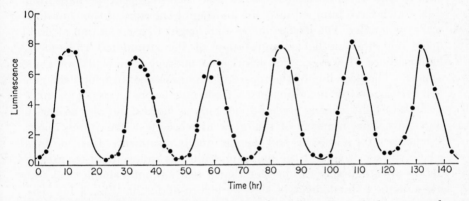

FIG. 3-10. The rhythm in stimulated luminescence in *Gonyaulax polyedra*, measured in LL (1000 lux). (Hastings and Sweeney, 1957.)

Rhythmicity can be detected in an isolated *Gonyaulax* cell, although not the rhythm in luminescence as yet.

When changes in the brightness of luminescence occur, this could be the result of alternation in the biochemistry through which luminescence is produced or in the mechanism for stimulating a luminescent flash or in both. We do not know as yet the precise nature of the changes which occur to bring about cycles. However, when care is taken to discharge as little light as possible during the extraction procedures, larger amounts of both luciferin and luciferase can be recovered from cells in the night phase than from those in the day phase. The difference in activity of the luciferin-luciferase system *in vitro* is about a factor of 2, while cell luminescence changes much more than this even in low intensity LL. Furthermore, Hastings and Bode (1962) found that exposing cells to bright light at midnight, a treatment which is known to reduce cell luminescence greatly, does not reduce the activity extractable from these cells. Thus a component of the rhythm in luminescence which depends on the stimulating mechanism is indicated.

Kelly and Katona (1966) discovered that dinoflagellate populations from the Eel Pond at Woods Hole which showed a luminescence rhythm like that of *Gonyaulax* also clearly possessed a rhythm in the sensitivity of the luminescence to light inhibition. It is not yet known whether such a rhythm in light inhibition is present in *Gonyaulax*.

Luminescence is not the only rhythmic function in *Gonyaulax*. Curiously enough, there is another rhythm which concerns luminescence. When a population of *Gonyaulax* in a tube is placed in front of a photomultiplier in darkness at the beginning of a night, but *not stimulated* to emit light, infrequent flashes none the less appear, perhaps from cells which have bumped into one another or the edge of the container while swimming. The flashes increase in height to a maximum at midnight and are another manifestation of the stimulated rhythm in luminescence. However, a few hours after midnight a new phenomenon appears. The base line of the record of luminescence, to which the pen had returned after recording every flash, now begins to rise. This must mean that light is being given off, not only as flashes, but also as a very dim glow. The intensity of this glow increases to a maximum about 4–6 hours after midnight, and then gradually disappears. In recordings where the sensitivity of the photo-multiplier is sufficient to record the details of this glow, it can be seen that it really is not a simple peak but has a distinct shoulder at or slightly before midnight (FIG. 3-11). The different time of the maximum distinguishes this 'glow rhythm' from that in stimulated luminescence. It can be shown that the glow rhythm also continues in LL when the light intensity is low and for a time in DD. In addition, temperature has exactly the same effect on period in both luminescence rhythms. The brightness of the glow, however, increases with a rise in temperature when the intensity of stimulated luminescence decreases. This glow rhythm has been used extensively by Hastings in experiments concerning rhythms, since he has devised a system for recording glow automatically, and consecutively in a number of samples. With this apparatus it is possible to measure the glow rhythm over a considerable time span in parallel in cell suspensions which have received different treatment.

When the photosynthesis of *Gonyaulax* is measured, either as oxygen production or carbon dioxide fixation, it too is found to be rhythmic. The rhythm continues in constant light of the same intensities that allow the observation of luminescent rhythms. The period is circadian. The maximum rate of photosynthesis occurs, as one would expect, in the middle of the day, and hence the phase of this rhythm is different

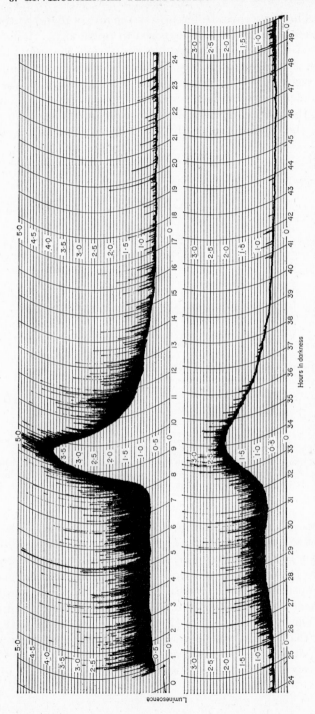

FIG. 3-11. A recording of the glow rhythm in *Gonyaulax polyedra* in DD. (Sweeney, unpublished.)

from that of luminescence. Differences in the rate of photosynthesis can only be detected if the cells are exposed to bright light. At low intensities, the actual rate of photosynthesis is always the same, although the rhythm is present and may be detected whenever samples are placed in bright light for a few minutes in the presence of $NaH^{14}CO_3$. It is the capacity for photosynthesis that is changing with time.

By using the Cartesian reference diver technique, it has been possible to measure the oxygen evolved by one isolated *Gonyaulax* cell in the

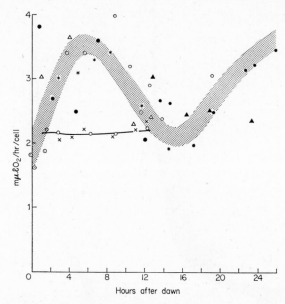

FIG. 3-12. Photosynthetic rhythm in a single *Gonyaulax polyedra* cell. Curve 1 (stippled): cell from LD 12 : $\overline{1}\overline{2}$, dawn at hour 0; curve 2 (continuous line): cell from bright LL. Note that the photosynthesis in treatment 2 was not rhythmic. (Redrawn from Sweeney, 1960.)

light. If this cell is taken from a culture on LD just after dawn, the rate of photosynthesis in saturating light can be measured throughout the day that follows. Photosynthesis in this cell changes in the same way as does the oxygen production of a population of cells, gradually increasing until noon, then decreasing (FIG. 3-12). A cell from continuous bright light however shows no changes in the rate of oxygen evolution. Thus we may conclude that light inhibits rhythmicity in each cell, rather than causing asynchrony among different cells of a population, each cell of which is rhythmic.

When oxygen production and carbon dioxide fixation are observed to vary, it seems reasonable to suppose that changes in photosynthetic biochemistry are directly responsible. Preliminary deductions concerning the nature of these changes can be made from the fact that the rhythm in photosynthesis in *Gonyaulax* represents a change in photosynthetic *capacity*, and cannot be observed at low light intensity. Changes in the efficiency of light gathering or light absorption cannot be involved, or photosynthesis would be rhythmic at low light intensity where light is limiting rate. Thus some dark reaction must be the rhythmic biochemical variable. This conclusion is confirmed by experiments with flashing light in which cells are exposed to a very bright but brief flash followed by a relatively long dark period. Under these conditions no rhythm is observed, as would be expected since all the dark reactions no matter how slow can be completed between flashes. Which dark reaction is changing in a rhythmic pattern? The rates of many of the dark reactions appear not to change with time in the cycle. However, cyclic changes *have* been found in the activity of ribulose diphosphate carboxylase, the enzyme responsible for incorporating CO_2 in the Calvin cycle (Sweeney, 1965).

Still one more rhythm with a circadian period can be observed in *Gonyaulax*. It has consistently been observed that almost all cell division occurs during 30 minutes in each 24 hours, when cultures are in LD; if the LD is 12 : 12, then this 30 minutes spans 'dawn'. An investigation of other LD schedules such as 7 : 7 shows that the dark-to-light transition is not the determining factor, since division takes place about 12 hours after the beginning of the dark period, even though this time is considerably after the beginning of the next light period. In continuous light of low intensity where other rhythms in *Gonyaulax* persist, very little cell division takes place and the *average* generation time may be as long as 6 days. However, those cells which are ready to divide do so only at the expected time in each 24 hours. Every day a few cells divide so that a circadian rhythm persisting in LL can be detected, a rhythm which cannot be due to synchrony of generation times, since such synchrony would yield a 6-day period.

When a cell is isolated in a capillary and observed frequently, the generation time for this cell can be determined directly. The most common generation time determined in this way for cells on LD was 48 hours, while 72- and 96-hour generation times were also found, but intermediate times did not occur. Stated in another way, the generation time of *Gonyaulax* is $n \times 24$ hours.

In single cells of *Gonyaulax*, and consequently in populations as well, there exists a temporal control over luminescence, photosynthesis and cell division, so that each process reaches a maximum then declines in an orderly sequence during each 24 hours. A single process, a biological clock, may control all these processes, at least there is no clear evidence as yet to suggest that these rhythms behave differently under a variety of conditions.

It may be worth noting that not all physiological processes in *Gonyaulax* are rhythmic by any means. The rate of respiration is constant with time, as is the incorporation of S^{35} into cellular components in general and P^{32} into DNA and RNA. No rhythm was found in the reactions of the photosynthetic electron transport system, such as the Hill reaction measured in whole cells in the presence of quinone, or the photo-reduction of dyes such as dichlorophenol indophenol in cell-free extracts, photo-phosphorylation or reduction of nicotinamide adenine dinucleotide phosphate.

A great many other rhythms have been investigated in sufficient detail to establish their circadian nature. Among the best known can be mentioned the rhythm in exudation from the root systems of severed *Helianthus* plants (Grossenbacher 1938, Hagan 1949, Vadia 1960, and MacDowall 1964); odour production in *Cestrum nocturnum* (Overland 1960); phototaxis in several species of diatoms isolated from the River Avon and investigated by Round and Palmer (1966); the growth patterns and spore discharge of *Neurospora* (Sussman *et al.* (1964), Sargent and Woodward (1957), Pittendrigh *et al.* (1959); and *Pilobolus* (Schmidle 1951, Uebelmesser 1954).

The most striking thing about circadian rhythms is their uniformity, in view of the great differences in organization between different organisms involved. What are the properties which all circadian rhythms have in common? The first is of course the circadian period itself, the free-running period which does not depart far from 24 hours *at any physiological temperature*, that is, the period is relatively temperature independent. This is rather remarkable, since temperature independence is achieved in spite of the fact that most if not all the biochemical reactions taking place within the cells in question show marked temperature dependence, usually doubling in rate with a 10° C rise in temperature. Although some animals which are rhythmic control their body temperature within narrow limits, most rhythmic cells have no such ability. Pittendrigh pointed out some time ago that temperature independence would be expected to have evolved in company with rhythmicity if timing a process had survival value, since a clock that runs faster on warmer days would be hopelessly inaccurate. Perhaps the evolution of temperature independence is less interesting than the observation that this temperature independence is not perfect. Some temperature dependence is still evident in almost all circadian rhythms, although the amount, even the direction, varies from organism to organism.

The discovery of constant conditions under which rhythms persist for long times made it possible to investigate the question of the dependence of the period on temperature, since the free-running period could now be measured with sufficient accuracy to detect small differences. The result of an experiment of this kind with *Gonyaulax* in LL at constant but different temperatures is shown in FIG. 3-13. After

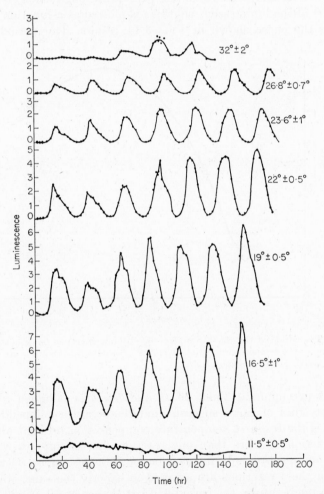

FIG. 3-13. The rhythm in stimulated luminescence in *Gonyaulax polyedra* in LL (1000 lux) at different temperatures ° C as indicated on each curve. Note that the period changes a little as the temperature is changed. (Hastings and Sweeney, 1957.)

seven days under constant conditions, the peaks of luminescence are only slightly displaced, and a clearly defined order to this displacement is visible: the period is longer the higher the temperature in the range between 16° C and 26° C, but the difference is small. If the ratio of the *frequency* of the oscillations at these two temperatures (the Q_{10}) is calculated, it is correspondingly less than 1,0·85 to be specific. The closeness of the Q_{10} to the value of 1 is an indication of the temperature independence of the system.

The effect of temperature on the period, although small, is real. If the period of the luminescent rhythms is plotted as a function of temperature, the curve shown in Fig. 3-14 results. The period passes

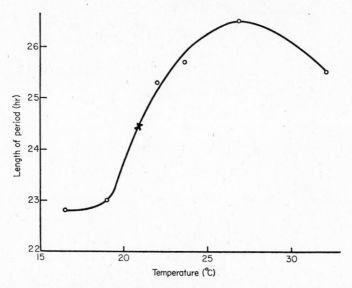

Fig. 3-14. The period of the rhythm in stimulated luminescence in *Gonyaulax polyedra* as a function of the temperature in LL (1000 lux). The data for this curve are shown in Fig. 3-13. (Hastings and Sweeney, 1957.)

through a maximum at about 26° C, and *decreases* at higher temperatures. This kind of curve suggests that temperature independence in this case is the result of temperature compensation brought about by the interaction of more than one reaction. Let us suppose that the result of one reaction is to increase period, of a second, to decrease it. If both of these reactions are affected in exactly the same way by a change in temperature, the period will not change at all, it will be perfectly temperature compensated. Imperfection in the matching between these two reactions such that the partner which lengthens period is accelerated slightly more than is the period-shortening reaction when temperature increases, would result in a small increase in period with rising temperature, as actually observed in *Gonyaulax*. If the period-lengthening reaction passed through an optimum at a somewhat lower temperature than its partner, then the curve for period length as a function of temperature would pass through a maximum, as observed.

The *Gonyaulax* rhythms are not the only ones in which the period is longer at higher than at lower temperatures. Bühnemann (1955c),

TABLE I

Temperature dependence in various plant circadian rhythms

Plant	Rhythm	Q_{10}*	Reference
Neurospora	Zonation in dim red light	1·03	Pittendrigh *et al.* (1959).
Pilobolus	Sporulation DD	1·3	Schmidle (1951).
		1·5	Uebelmesser (1954).
Gonyaulax	Luminescence stimulated cultures in dim light	0·85	Hastings and Sweeney (1957).
	Luminescent glow DD	0·9	Sweeney and Hastings (1958). Hastings and Sweeney (1959).
	Cell division dim light	0·85	Sweeney and Hastings (1958).
Oedogonium	Sporulation LL	0·8	Bühnemann (1955).
Euglena	Phototaxis DD and test light	1·01–1·1	Bruce and Pittendrigh (1956).
Helianthus	Exudation	1·1	Grossenbacher (1939).
Avena coleoptile	Growth DD	1	Ball and Dyke (1954).
Phaseolus	Sleep movement	1·3	Bünning (1931).
		1·01	Leinweber (1956).

$$* \quad Q_{10} \frac{\text{frequency } t}{\text{frequency}^{t-10}} \quad \text{or} \quad \frac{1/\text{period}^{t}}{1/\text{period}^{t-10}}$$

observed the same phenomenon in the sporulation rhythm in *Oedogonium*. The Q_{10} for this rhythm is 0·8. However, most rhythms run faster, i.e. have shorter periods at higher temperatures, (Table I). Whatever the direction of change, in all circadian rhythms this change is small, and consequently the values for Q_{10} are much less than 2.

The length of the period of circadian rhythms may be altered to a small extent by changing the light intensity of a constant light regime. The intensity can only be altered within the range where rhythmicity persists of course, and this limitation prevents the investigation of a large range of light intensities. Furthermore, indirect effects arising from changes in photosynthesis or photoinhibition of the process by which the rhythm is being measured must be taken into consideration.

Aschoff (1960), has gathered together the evidence concerning the effect of light intensity on animal rhythms, and has proposed an imaginative generalization. He finds that in diurnally-active animals the period decreases as light intensity is raised, while the reverse is true in nocturnal creatures. He points out that other factors which increase activity, raising the temperature for example, also generally shorten the period of rhythms. This correlation between activity and shorter periods has come to be called 'Aschoff's rule' and fits what is known about the effects of light and temperature on the period of circadian rhythms in activity in the diurnal finches, lizards and *Drosophila* and in several nocturnal mice and the hamster. It is not so satisfying when applied to rhythms observed in plants. For example, since the rhythm in luminescence in *Gonyaulax* runs with shorter periods the brighter the light, this organism would be diurnal according to Aschoff's rule. However, the intensity of luminescence becomes reduced as the light intensity is increased, and the period is longer at higher temperatures. These observations run counter to Aschoff's generalization. There seems no real basis for assigning any meaning to 'diurnal' or 'nocturnal' in plant systems where one function ordinarily associated with day like photosynthesis, and another typically 'nocturnal' like luminescence may both show circadian rhythmicity with exactly the same response to changes in temperature and perhaps light intensity as well. Where an effect of light intensity on period can be detected, the change in period is always quite small, a few hours at most.

Bright white light inhibits rhythmicity in *Gonyaulax*. Is this a universal characteristic of circadian rhythms? Like *Gonyaulax*, *Euglena* is arhythmic in bright continuous light, as in the mould *Pilobolus*. Even light of low intensity inhibits rhythmicity in *Kalanchoë* and *Avena*. On the other hand, some plant rhythms are only found in LL and it is not clear from the original papers what the maximum light intensity for the persistence of rhythmicity may be. The question is complicated since the spectral distribution of the light is known to be important in several cases. For example, in *Phaseolus* rhythms persist for a long time in red LL (600–650 mμ) but disappear at once in far-red LL (700 mμ). Thus the intensity tolerated for the continuance of rhythmicity depends on the relative content of far-red light. Bünning's laboratory uses LL 80–500 lux. Hoshizaki and Hamner (1964) detect rhythms at much higher intensities. Another complicating factor is the effect of pretreatment. Wilkins (1959, 1960) has documented this in *Bryophyllum*. The rhythm in CO_2 output in leaves of this plant is ordinarily measured in

DD. After some time in darkness, 25 f.c. light is sufficient to stop rhymicity. However, if leaves had been held at higher light intensities, then they became rhythmic on lowering the light intensity to 25 f.c. After investigating the effects of a number of different combinations of light intensity he concluded (FIG. 3-15) that the *difference* in light intensity rather than the intensity itself determined whether or not a rhythm would be observed.

Arhythmia in a population of cells could result from stopping the oscillation, uncoupling the process in question from the oscillator, or

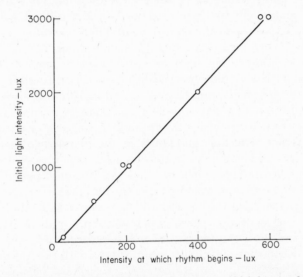

FIG. 3-15. The relationship between the light intensity to which leaves of *Bryophyllum fedtschenkoi* have been exposed and the intensity at the rhythm in CO_2 output can be observed in LL. (Wilkins, 1960. *Cold Spring Harb. quant. Biol.* **25**, 115–129.)

from asynchrony between cells all of which were still rhythmic. In *Gonyaulax*, photosynthesis and cell division in single cells are not rhythmic in bright LL. Thus asynchrony is not responsible for the lack of rhythmicity in populations under these conditions. The fact that such different processes as photosynthesis and cell division fail to be rhythmic in single cells in bright light makes it seem unlikely that the coupling between these processes and the oscillator, which would presumably be quite different, would both be inhibited by light at precisely the same light intensity. Therefore a stopping of the clock by bright light seems the favoured hypothesis. This may not be true in all systems showing inhibition of rhythmicity by strong light. For example, the composite flowers of Chicory grown in bright LL do not open normally

(Todt, 1962). A few days in LL is sufficient to upset flower opening, so that some florets of a flower open at one time, some at another. If the LL regime is continued for a longer time, flower opening is increasingly asynchronous. Since the florets open only once, a rhythm cannot be detected in a single flower, but the results suggest asynchrony rather than the uncoupling or stopping of the clock.

The inhibition of rhythmicity in *Gonyaulax* by bright LL makes certain experiments possible which could not have been done, had loss of rhythmicity been due to asynchrony between member cells of a population. These are experiments to answer the question, 'Is the circadian period an innate character programmed in the cell's DNA or is it 'learned' by exposure to the environmental daily periodicity?' Populations can be grown for many generations in LL. During this time they will not be rhythmic. The light intensity can then be lowered or the cells can be darkened and observed in DD. Such a procedure provides the cells with no information from the environment concerning period. The result of this experiment was clear: rhythmicity was reinitiated just by lowering the light intensity. No training was necessary to relearn the circadian period. The first maximum in the luminescence was found 6 hours after darkening, as if this transfer was interpreted as the beginning of night. The conclusion that the circadian period is innate and not learned was supported by experiments where cells were cultured for a long time in light-dark cycles which were not 24 hours long. Such *Gonyaulax* cells showed a *circadian* period at once on being transferred to LL, of the proper intensity. Semon's experiments with different cycles of LD had shown a similar result in 1905 with the leaf movement rhythm in *Acacia*.

Light-dark cycles are not necessary to initiate a rhythm in the leaf movement in *Phaseolus*. In this case, arhythmic plants are obtained by germinating the beans in bright LL. One dark period interposed in this LL can initiate a leaf rhythm provided darkness is at least 9 hours long (Wassermann, 1959). Changes in temperature can also sometimes initiate a rhythm which has been lost in LL. Such an effect probably accounts for the observation of Hillman that tomato plants will not grow in a strictly constant light and temperature regime, but will prosper if a temperature change occurs from time to time.

While most rhythmic systems behave in a manner consistent with the inheritance of the information regarding period, there is one notable exception. In the green alga *Hydrodictyon*, Pirson and his collaborators found that growth rate, photosynthesis and respiration were periodic, and that this periodicity persisted for several cycles when the alga was transferred to constant light and temperature. However, in constant

conditions, the period may not be circadian, but is the same as that of
the *previous* LD, even if this was LD 6 : 6̄ or LD 10½ : 7. *Hydrodictyon*
thus appears to 'learn' different periods. Since rhythmicity disappears
after only a few cycles, it is not possible completely to eliminate the
possibility that the period observed in LL was actually the result of
events occurring while the alga was still under LD conditions. Some
precedent for such a lag may be found in the sporulation rhythm of
Pilobolus studied by Uebelmesser (1954). When this mould is transferred from LD 6 : 6̄ to DD, three short periods are observed before the

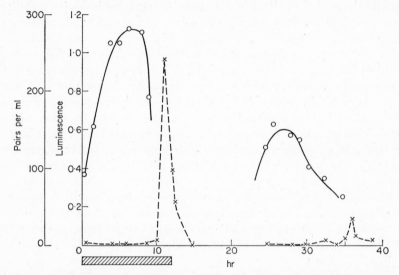

FIG. 3-16. The rhythm in luminescence (circles) and cell division (crosses) in *Gonyaulax*
polyedra just before and just after the beginning of LL (1000 lux) following LD 12 : 1̄2̄,
18° C. The black bar below the abscissa represents the last dark period. Note the change
in the relative position of the two maxima in the first cycle after the beginning of continuous light. (Sweeney, unpublished.)

circadian free-running period is established. In the cell division rhythm
of *Gonyaulax* also, the first period after the beginning of LL is 24 hours
if the previous LD was 12 : 1̄2̄, even at a temperature where the eventual
free-running period will be much shorter. (FIG. 3-16). Luminescence,
on the other hand, immediately takes up the short free-running period
after cells are transferred to LL. These results can be interpreted quite
simply if one assumes that cell division is determined some 12–18
hours prior to actual mitosis, while luminescence does not show such
a lag.

    It might be well to emphasize once more that there is not one free-
running period in any circadian rhythm, but a number of somewhat

different periods are possible in constant conditions depending on the temperature and the light intensity. Usually these values cluster on either side of 24 hours, but this is not necessarily always the case, some rhythms usually showing periods a little shorter than 24 hours like *Euglena* phototaxis, and some a little longer, like *Phaseolus* leaf movement.

PHASE DETERMINATION

While the period of circadian rhythms is remarkable for its stability, the phase of all such rhythms is labile; it can easily be changed. The comparison has often been drawn between rhythms and mechanical clocks which run with a constant speed but possess hands, the position of which can easily be reset. All circadian rhythms are conspicuously similar in the manner in which the phase can be changed or reset. One of the first experimental findings regarding the determination of phase in circadian rhythms was that phase could be reversed by a reversal of the position of light and darkness in the environment. Rhythms adjust very quickly to artificial 'day' at night, fortunately for those making measurements of these phenomena. The phase of all circadian rhythms appears to be completely labile in the sense that maxima may be produced at any desired solar time, simply by exposing the cells in question to the appropriate LD regime. The rhythm will continue in the phase set by the last LD after transfer to constant conditions, showing the period dictated by the particular temperature and light intensity used.

The phase of a circadian rhythm is completely predictable both in LD and in LL, provided that the phase response curve for the rhythm in question has been determined. The shape of the phase response curves of circadian rhythms is characteristic, and represents one of the most interesting features that these rhythms have in common. In the 1950's, when a number of rhythms were first being examined in detail in this respect, it was discovered that resetting could be brought about by a single exposure to visible light, in place of a changed LD cycle. This exposure to light would only be effective, if it were placed in the part of the circadian cycle characteristic of darkness. Furthermore, all parts of the dark phase were not equally sensitive to resetting by light. Light exposures in the middle of this phase were far more effective in shifting phase. Not only was there a difference in the amount of phase shift which depended on the time of the light exposure but the *direction* of the shift was also dependent on when in the cycle the light exposure was administered. Another way of expressing these observations quantitatively is to plot a phase response curve, in which the amount of reset on the ordinate is shown as a function of the phase of the cycle when it was given (FIG. 3-17). If the first maximum in the rhythm under study occurred later following a light signal than in the untreated control, in other words when there was a phase delay, the sign of the phase change was plotted as positive in the earlier papers, while phase advances, i.e. earlier maxima, were considered negative. However, consultation with physicists disclosed that the opposite sign is given to phase shifts in

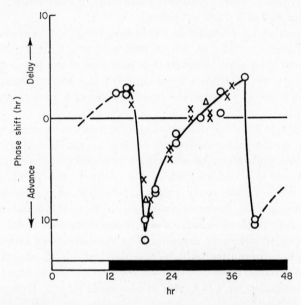

FIG. 3-17. A phase-response curve for the rhythms in *Gonyaulax polyedra*. Data from various experiments in which cells, otherwise in LL, were given a single 3-hour light exposure; circles are from the glow rhythm, crosses are from the rhythm in stimulated luminescence, triangles are from the rhythm in cell division. (0 time is taken as the beginning of the last light period of the LD preceding DD.) (Sweeney, unpublished, and Hastings and Sweeney, 1960.)

physical oscillations. Thus it was decided to conform to the already established convention and call a phase advance positive in rhythm papers subsequent to the Summer School on Circadian Clocks at Feldafing, in the summer of 1963. In any case, the phase response curve has the form of a derivative in all circadian rhythms examined. The slope of the curve differs from organism to organism, as does the relative amounts of advances and delays.

Recently, it has become apparent that an exposure to darkness in an LL regime can also cause a shift in phase, provided that it occurs at the right time in the cycle. The maximum sensitivity to darkness is in the day phase, and again both delays and advances in phase can be produced, depending on when the dark exposure occurs. The phase response curve for darkness appears to be the mirror image of that for light. Experiments in which *Gonyaulax* was transferred to darkness from low intensity LL at different times in the cycle showed that a single stepdown in light intensity during the day phase resets the phase of the glow rhythm. Thus the effect must truly be attributable to the lack of light, rather than to some changed sensitivity to light when it comes on again at the end of a dark exposure.

Experiments with *Gonyaulax* (Sweeney, 1963) and *Kalanchoë* (Bünning and Zimmer, 1962), in which two resetting light exposures were given in short succession, showed that resetting of the phase takes place for all intents and purposes immediately; the second light exposure resetting as if the rhythm were already in the new phase as a result of the first.

Pittendrigh (1954) has discussed the interesting consequences of the differences in the phase response curves to visible light in diurnal, nocturnal and photoperiodic animals. The phase response curve for the Flying Squirrel was determined by De Coursey. She pointed out that the phase delay in the early part of the night phase would serve to prevent this animal from exposing itself repeatedly during the increasingly long days of spring, while phase advances would serve the same function in the early morning. The significance of phase response curves in photo-periodism will be discussed in a later chapter.

It has been clear from the time of the early experiments that phase could be controlled by LD cycles of abnormal length. Different organisms vary in their ability to entrain to cycles very different from LD 12: $\overline{12}$, especially to those where the sum of the light and the dark period is not 24 hours. The behaviour of a particular organism can be predicted from a knowledge of the phase response curve of this organism and the dose necessary to bring about various responses. A number of experiments have shown that in general plants can be entrained to shorter cycles than can animals. For example, the luminescence in *Gonyaulax* cell suspensions kept on LD 6: $\overline{6}$ increases throughout the short dark period, and falls rapidly at the beginning of the light period. The light has come on at the middle of the night phase, the time when the cells are most sensitive to resetting by light. Resetting occurs so that 6 hours later at the end of the light period luminescence begins another night phase.

Another consequence of the characteristics of the phase response curves is that in light-dark cycles such as LD 16: $\overline{8}$ where the day is long compared to the time in darkness, the night phase will not begin after 12 hours light as it normally would do on LD 12: $\bar{1}\bar{2}$ This happens because light at the beginning of the 'night' shifts phase to a later time i.e. negatively. Conversely, if the extra light coincides with the last part of the night phase, the shift will be to an earlier beginning of the next phase. The whole effect will be to crowd events characteristic of the night phase into the dark part of the cycle. Such mechanisms may

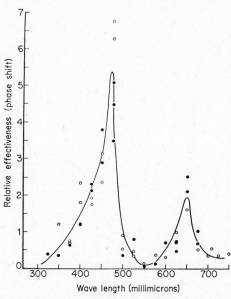

FIG. 3-18. The action spectrum for resetting phase in the rhythm of stimulated luminescence in *Gonyaulax polyedra*. Cells otherwise in DD were exposed to light of different wave lengths for 3 hours at the middle of the night phase. (Hastings and Sweeney, 1960.)

be important in nature where large variations in day length with season occur.

When monochromatic light is tested for its power to reset rhythms, a variety of action spectra are found. This suggests that the pigments responsible for detecting light are different in different rhythmic systems, and may not even be integral parts of the basic oscillators involved. With respect to the rhythm in luminescence in *Gonyaulax*, the action spectrum for resetting phase (FIG. 3-18) shows two sharp maxima, one at 475 mμ and the other at 650 mμ. Thus the blue maximum is at a longer wave length than that for photosynthesis (about 444mμ) while at the red end of the spectrum, the maximum falls at too short a wave length for chlorophyll *a* (650 mμ instead of 680 mμ). We do not know how much the action spectrum is influenced by the presence of inactive pigments which shade the active site for resetting. As it is, the action spectrum does not fit the absorption spectrum of any pigments that we know are present in *Gonyaulax* cells. It perhaps resembles the absorption spectrum of chlorophyll *b*, but this chlorophyll has not been found in dinoflagellates. In these organisms, it is replaced by chlorophyll *c*, which has only two small absorption peaks in the red end of the spectrum. Thus the action spectrum for resetting phase has

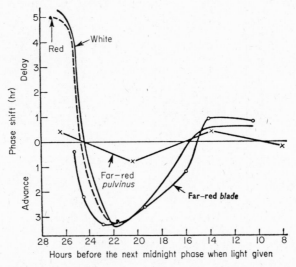

FIG. 3-19. The phase-response curves for the sleep movement rhythm in *Phaseolus multiflorus* exposed to white, red or far-red light. Time is given as hours before the next midnight phase of the unirradiated controls. (Bünning and Moser, 1966.)

not so far offered any great insights. We do know, however, from experiments where red light followed exposure to far red light, or vice versa, or both wave lengths were present, that there is no evidence for red-far red reversal in resetting, and hence no suggestion that a phytochrome-like pigment is active.

No action spectrum for phase shifting is available for any other plant rhythm. However, some knowledge of the general region of the spectrum which is effective has been gained for several other systems. In *Phaseolus* (FIG. 3-19) Bünning and Moser (1966), showed that a phase response curve identical to that in white light may be obtained by irradiating the pulvinus with red light with a spectral distribution between 600–720 mμ and maximum emission at 660 mμ. Far-red light

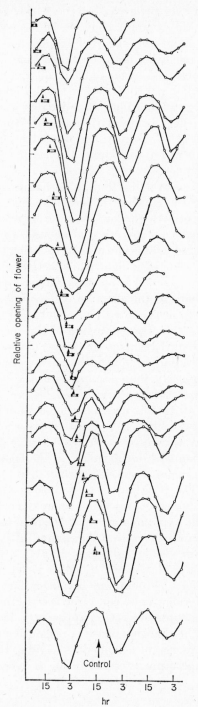

Relative opening of flower

Control

15   3   15   3   15   3

hr

Fig. 3-20. The phase shift induced in the rhythm of petal movement in *Kalanchoë blossfeldiana* by exposure to red light (emission maximum 660 mμ) for 2 hours, indicated by the bar under each curve. Plants were otherwise in DD. (Zimmer, 1962. *Planta*, **58**, 283–300.)

on the pulvinus caused very little phase shift; if this light fell on the leaf blade however, phase advances could be observed, but no phase delays. It seems likely that the effect of far-red light is essentially different from that of red light, both for this reason and because no rhythm at all can be observed in far-red LL, although the *Phaseolus* rhythm persists very well in red LL (Lörcher 1958). Light sources which yielded predominantly green and blue wave lengths were less effective in phase shifting, any effect being attributable to contaminating red light. The intensities and exposure times needed for resetting argue against the involvement of phytochrome, nor has red-far-red reversal been demonstrated.

Red light of the same spectral distribution as that used with *Phaseolus* is effective in resetting the rhythm of petal movement in *Kalanchoë* (FIG. 3-20). The *Avena* growth rhythm can also be reset by red light. Other colours of light were not investigated in either rhythm, however.

In *Bryophyllum* also, red light is effective in resetting the phase of the rhythm in CO_2 output, while blue light is ineffective, green light being intermediate in its effect. Broad-band pass filters were used, so that no conclusions with regard to the effective pigment can be made.

In fungi, as distinct from these red-sensitive higher plant rhythms, blue light appears to be the effective region of the spectrum as far as phase shifting of circadian rhythms is concerned (Sargent, 1967).

The rhythm in luminescence in *Gonyaulax* can be reset by ultra-violet as well as by visible light. In these studies Sweeney used a sterilamp which emits principally at the mercury line at 254 mμ. An exposure of only a few minutes is all that is necessary for a complete reversal of phase, rather than hours as with visible light. There is another interesting difference between the reset caused by visible light and that resulting from irradiation with ultra-violet: in the latter all resets are advances, the next maximum occurring earlier than it would have in the absence of light. The possibilities of confusing very long delays with advances was eliminated by observing where the maxima fell when less than optimum light exposures were used. The sensitivity to reset by ultra-violet varies cyclically with a maximum in the middle of the dark phase, just as with visible light (FIG. 3-21). Visible light does not appear to reverse the phase shift by ultra-violet as it does some effects of ultra-violet light, chromosome breakage, for example. Ehret (1960) finds that ultra-violet light can reset the rhythm of mating in *Paramecium*, but here photoreactivation apparently does occur.

In the absence of light changes, phase is responsive to temperature cycles. Even in the presence of a light-dark regime, the effects of an opposing high-low temperature alternation can be detected. The direction of phase changes indicates that a rise in temperature is interpreted as day and a fall as night. The question of the effect of single temperature pulses has been the motivation for a number of studies, in particular by Bünning and his students. They found that the leaf-movement

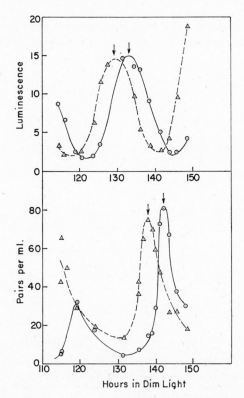

FIG. 3-21. The phase shift of the rhythm of stimulated luminescence in *Gonyaulax polyedra* in response to ultra-violet light (2 minutes, 150 ergs cm^{-2} sec^{-1} at time 0). The upper curves show the rhythm in stimulated luminescence; the lower curves, the rhythm in cell division; dashed curves, irradiated cells; solid curves, unirradiated controls. (Sweeney, 1963.)

rhythm in *Phaseolus* is insensitive to short exposures to either warmer or cooler temperatures during a part of each cycle, while during another part, marked changes in the oscillation result from such treatment. For example, Moser found that when the temperature was raised from 20° C to 28° C for 4 hours in an LL regime, there was no reset, provided the temperature was raised during the day phase. Phase delays occurred when high temperature coincided with the middle of the night phase.

Later in the night phase a phase advance was produced, dawn being the most sensitive part of the cycle (FIG. 3-22). The derivative shape of the curve depicting the phase change in response to high temperature at different parts of the cycle is reminiscent of the phase response curve for light. In Moser's experiments, the final phase was not attained at once following the exposure to high temperature, but required several transient cycles. The high temperature need not be given as a short pulse to bring about a phase change; raising the temperature to

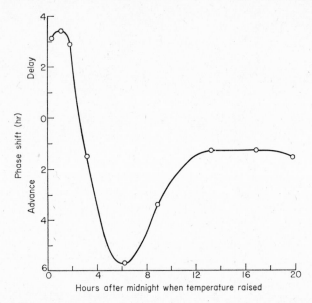

FIG. 3-22. The phase response curve of the rhythm in leaf movement in *Phaseolus multiflorus* to raising the temperature from 20° C to 28° C for 4 hours at different times in the cycle in LL. (0 time is taken as the middle of the night phase when leaves are in the lowest position.) (Moser, 1962.)

28° C in an LL schedule will also cause phase shifts as predicted from the experiments with 4-hour pulses. The transients resulting from this effect must be taken into account in planning experiments where measurements of period in LL at different temperatures are to be made. This was not known at the time of Bünning's earlier determinations of the temperature dependence of the period of the *Phaseolus* rhythm. His failure to discard the first few periods after a temperature increase resulted in a temperature dependence somewhat greater than that determined in later experiments where this source of error was taken into consideration (Leinweber (1961); $Q_{10} = 1·01$ as compared with Bünning (1931); $Q_{10} = 1·3$).

Short exposures to low temperature, below 10° C, also may have different effects on rhythms at different phases. Bünning has paid much attention to this phenomenon because the insensitivity of rhythms to cold during part of a cycle could provide evidence that energy yielding processes are not required for clock maintenance during this phase. Relaxation oscillations might be expected not to require energy during the relaxation phase. Bünning has concluded that the circadian clock

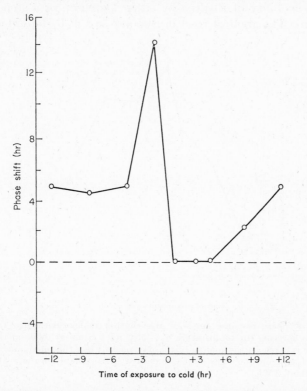

Fig. 3-23. The phase response curve of the rhythm in leaf movement in *Phaseolus multiflorus* to lowering the temperature from 20° C to 4·5° C for 5 hours at different times in the cycle in LL (0 time is taken as the middle of the night phase when leaves are in the lowest position, and phase shifts are calculated with respect to the control in LL, and plotted at the time when the cold exposure ended). (Wagner, 1963.)

is an oscillator of this type. In *Phaseolus*, such an interpretation is favoured by the observation that the greatest sensitivity to cold (5 hour treatments), is found at the time of the midnight minimum leaf opening, or just before, while later in the night phase cold is without much effect (Fig. 3-23). Furthermore, exposures to 5° C for more than 20 hours seems to stop the clock at a point 8 hours before midnight, no matter how long the cold treatment may be, as if this were the point where the tension phase which required energy would normally begin. The sporulation rhythm in *Oedogonium* responded to short cold treatments in a differential manner also (Ruddat). The time of maximum sensitivity to cold treatments is just after the sporulation maximum (FIG. 3-24), or, unlike *Phaseolus*, during midday phase.

Brinkmann recently published curves for the phase change of the rhythm in mobility of *Euglena* by either a step-up or a step-down in temperature. The greatest reset in response to a 5° C step-up occurs at

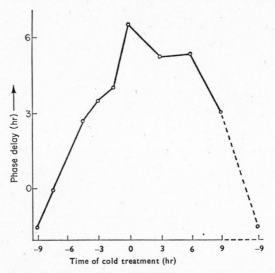

FIG. 3-24. The phase response curve of the rhythm in sporulation in *Oedogonium cardiacum* to lowering the temperature to −5° C for 6 hours at different times in the cycle (0 time is taken as the middle of the day phase when the maximum number of spores are released and the phase shifts are plotted at the time when the cold exposure ended). (Ruddat, 1961.)

the middle of the night phase (FIG. 3-25). A step-down of the same magnitude was effective only if administered within an hour or so of dawn (although the middle of the day phase is insensitive to both step-up and step-down in temperature). These curves are certainly not identical in shape and can hardly be cited as evidence for a temperature-sensitive and a temperature-insensitive half cycle.

Wilkins has shown that the rhythm of CO_2 fixation in *Bryophyllum* is stopped by a 4-hour exposure to low temperature at or near midday phase but unaffected during about 4 hours of the night phase when CO_2 output is low (CO_2 fixation is highest).

If there exists a part of the cycle of circadian rhythms during which energy is not utilized, then it would be predicted that this part of the cycle would be insensitive to anaerobic conditions, as well as to low temperature. Rhythms in *Phaseolus* and *Kalanchoë* were examined with this question in mind by Bünning, Kurras and Vielhaben (1965).

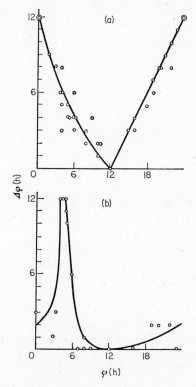

FIG. 3-25. Phase response curves of the rhythm in mobility and phototaxis in autotrophic cultures of *Euglena gracilis* to raising (curve a) or lowering (curve b) the temperature by 5° C in the temperature range where the period is unaffected by change in temperature (0 time is taken as the middle of the night phase when mobility is minimal). (Brinkmann, 1966. *Planta*, **70**, 344–389.)

Flower opening in *Kalanchoë* was unaffected if an atmosphere of N_2 was supplied for 4 hours at the end of the day and beginning of the night phase, while phase delays of up to 4 hours occurred when the exposure was during the dawn phase (FIG. 3-26). Very similar results were also obtained with the bean leaf rhythm, with the same phase sensitivity. However, the phase sensitive to anaerobiosis was *not* that sensitive to cold, reaffirmed as the midnight phase in *Phaseolus*. Treatment with the respiratory inhibitors, cyanide and dinitrophenol, produced their greatest effect on phase when present from the eighth hour after midnight as did anaerobic conditions (Keller, 1960). Cyanide delayed, but dinitrophenol *advanced* phase.

The growth rhythm in *Avena* which is initiated when plants are transferred from red light to darkness can be reset by short exposures to anaerobic conditions. If treatment was begun 12 hours after the beginning of darkness and continued for 2, 4 or 6 hours over the time of maximum growth in controls, the delay in the next maximum was about equal to the time in N_2. There was no time however when

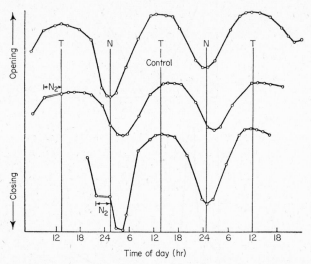

FIG. 3-26. The phase shift of the rhythm of petal movement in flowers of *Kalanchoë blossfeldiana* in DD in response to 4 hours in an atmosphere of N_2 (0 time is taken as the middle of the night phase of the control). (Bünning, Kurras and Vielhaben, 1965.)

anaerobic conditions were without effect on the rhythm. Delays seemed to be bigger when less time passed between the end of the N_2 treatment and the next maximum in growth. After anaerobiosis, a speeding up of time keeping, manifest in the second and third peak after treatment, seems to take place in *Avena*. Delays may be reduced by this compensating effect. Partial anaerobiosis brings about delays proportional to the reduction in the oxygen concentration, and no time especially sensitive to low oxygen concentrations was detected. Since there is no part of the cycle when lack or severe reduction in oxygen pressure does not delay the rhythm in *Avena*, metabolic energy must be required for the oscillator in all phases. Therefore the idea that this oscillator is of the relaxation type is not supported.

The similarity in the general form of the response of a wide variety of plant and animal systems to phase shifting stimuli is really quite remarkable. The question whether this reflects a similarity in the nature of the basic oscillator responsible for rhythms cannot be answered without further information. One of the greatest frustrations for those investigating the nature of the circadian oscillations is their inability to measure the oscillation directly, even to detect its general form, as distinct from the rhythms which it directs and which are clearly different in shape from each other. No doubt none of them reflect the basic oscillator in every detail. Period and phase changes are obviously the only possible indications of changes in the basic oscillator. Hence particular attention must be focused on these properties, especially the sensitivity to phase shifting by light, since all rhythms seem to show maxima in resetting in the middle of the night phase and the shape of the phase response curves in all organisms are so similar. Here perhaps the basic oscillator reveals itself.

REFERENCES

Aschoff, J. (1960). Exogenous and endogenous components in circadian rhythms. *Cold Spring Harb. Symp. quant. Biol.* **25**, 11–28.

Ball, N. G., and Dyke, I. J. (1954). An endogenous 24-hour rhythm in the growth of the *Avena* Coleoptile. *J. exp. Bot.* **5**, 421–433.

Bode, V. C., DeSa, R., and Hastings, J. W. (1963). Daily rhythm of luciferin activity in *Gonyaulax polyedra*. *Science*, N.Y. **141**, 913–915.

Brinkmann, K. (1966). Temperature inflüsse auf die circadiane Rhythmik von *Euglena gracilis* bei Mixotrophie und Autotrophie. *Planta* **70**, 344–389.

Brouwer, G. (1926). De periodieke Bewegingen van de primaire Bladeren bij *Canavalia ensiformis*. 110 p. H. J. Paris, Amsterdam.

Bruce, V. G., and Pittendrigh, C. S. (1956). Temperature independence in a unicellular 'clock'. *Proc. natn. Acad. Sci. U.S.A.* **42**, 676–682.

Bühnemann, F. (1955a). Die rhythmische Sporenbildung von *Oedogonium cardiacum* Wittr. *Biol. Zbl.* **74**, 1–54.

Bühnemann, F. (1955b). Das endodiurnale System der *Oedogonium*-Zelle. II. Der Einfluss von Stoffwechselgiften und anderen Wirkstoffen. *Biol. Zbl.* **74**, 691–705.

Bühnemann, F. (1955c). Das endodiurnale System der *Oedogonium*-Zelle. III. Über den Temperatureinfluss. *Z. Naturf.* **10b**, 305–310.

Bühnemann, F. (1955d). Das endodiurnale System der *Oedogonium*-Zelle. IV. Die Wirkung verschiedener Spektralbereiche auf die Sporulations-und Mitoserhythmik. *Planta* **46**, 227–255.

Bünning, E. (1931). Untersuchungen über die autonomen tagesperiodischen Bewegungen der Primarblatter von *Phaseolus multiflorus*. *Jb. wiss. Bot.* **25**, 439–480.

Bünning, E. (1956). Endogene Aktivitätsrhythmen. *In* Handbuch der Pflanzenphysiologie, vol. II. pp. 878–907, (ed. W. Ruhland). Springer, Berlin.

Bünning, E. (1960). Opening address: Biological Clocks. *Cold Spring Harb. Symp. quant. Biol.* **25**, 1–9.

Bünning, E. (1962). Mechanism in circadian rhythms: functional and pathological changes resulting from beats and from rhythm abnormalities. *Ann. N. Y. Acad. Sci.* **98**, 901–915.

Bünning, E. (1967). The physiological clock. Edition 2. 167 pp. Springer-Verlag, New York.

Bünning, E., Kurras, S., and Vielhaben, V. (1965). Phasenverschiebungen der endogenen Tagesrhythmik durch Reduktion der Atmung. *Planta* **64**, 291–300.

Bünning, E., and Moser, I. (1966). Response-kurven bei der circadianen Rhythmik von *Phaseolus*. *Planta* **69**, 101–110.

Bünning, E., and Tazawa, M. (1957). Über den Temperatureinfluss auf die endogene Tagesrhythmik bei *Phaseolus*. *Planta* **50**, 107–121.

Bünning, E., and Zimmer, R. (1962). Zur Deutung der Phasenverschiebungen und 'transients' nach exogener Störung endogener Rhythmen. *Planta* **59**, 1–14.

Bünsow, R. C. (1960). The circadian rhythm of photoperiodic responsiveness in Kalanchoë. *Cold Spring Harb. Symp. quant. Biol.* **25**, 257–260.

Ehret, C. F. (1960). Action spectra and nucleic acid metabolism in circadian rhythms at the cellular level. *Cold Spring Harb. Symp. quant. Biol.* **25**, 149–158.

Engelmann, W. (1960). Endogene Rhythmik und photoperiodische Blühinduktion bei Kalanchoë. *Planta* **55**, 496–511.

Grossenbacher, K. A. (1938). Diurnal fluctuations in root pressure. *Pl. Physiol. Lancaster*, **13**, 669–676.

Hagan, R. M. (1949). Autonomic diurnal cycles in the water relations of non-exuding detopped root systems. *Pl. Physiol. Lancaster*, **24**, 441–454.

Halberg, F., Halberg, E., Barnum, C. P. and Bittner, J. J. (1959). Physiologic 24-hour periodicity in human beings and mice, the lighting regime and daily routine. *In* Photoperiodism and related processes in plants and animals. pp. 803–878 (ed. R. B. Withrow), AAAS, Washington D.C.

Hastings, J. W. (1959). Unicellular clocks. *A. Rev. Microbiol.* **13**, 297–312.

Hastings, J. W. (1964). The role of light in persistent daily rhythms. *In* Photophysiology, vol. I, pp. 333–361 (ed. A. Geise). Academic Press, New York.

Hastings, J. W., Astrachan, L. and Sweeney, B. M. (1960). A persistent daily rhythm in photosynthesis. *J. gen. Physiol.* **45**, 69–76.

Hastings, J. W., and Bode, V. C. (1962). Biochemistry of rhythmic systems. *Ann. N.Y. Acad. Sci.* **98**, 876–889.

Hastings, J. W. and Keynan, A. (1965). Molecular aspects of circadian systems. *In* Circadian Clocks. pp. 167–182 (ed. J. Aschoff). North-Holland Publishing Co., Amsterdam.

Hastings, J. W., and Sweeney, B. M. (1957). On the mechanism of temperature independence in a biological clock. *Proc. natn. Acad. Sci. U.S.A.* **43**, 804–811.

Hastings, J. W., and Sweeney, B. M. (1958). A persistent diurnal rhythm of luminescence in *Gonyaulax polyedra*. *Biol. Bull. mar. biol. Lab., Wood's Hole*, **115**, 440–458.

Hastings, J. W., and Sweeney, B. M. (1960). The action spectrum for shifting the phase of the rhythm of luminescence in *Gonyaulax polyedra*. *J. gen. Physiol.* **43**, 697–706.

Hastings, J. W., and Sweeney, B. M. (1964). Phased cell division in the marine dinoflagellates. *In* Synchrony of Cell Division and Growth. pp. 307–321 (ed. E. Zeuthen). John Wiley and Sons, London.

Hellebust, J. A., Terborgh, J., and McLeod, G. C. (1968). The photosynthetic rhythm of *Acetabularia crenulata*. II. Measurements of photo-assimulation of carbon dioxide and the activities of enzymes of the reductive pentose cycle. *Biol. Bull. mar. biol. Lab., Wood's Hole*. (In press.)

Hillman, W. S. (1956). Injury of tomato plants by continuous light and unfavorable photoperiodic cycles. *Am. J. Bot.* **43**, 89–96.

Hoshizaki, T., and Hamner, K. C. (1964). Circadian leaf movements: persistence in bean plants grown in high-intensity light. *Science, N.Y.*, **144**, 1240–1241.

Karvé, A., Engelmann, W., and Schoser, G. (1961). Initiation of rhythmic petal movements in *Kalanchoë blossfeldiana* by transfer from continuous darkness to continuous light or vice versa. *Planta* **56**, 700–711.

Keller, S. (1960). Über die Wirkung chemischer Faktoren auf die tagesperiodischen Blattbewegungen von *Phaseolus multiflorus*. *Z. Bot.* **48**, 32–57.

Kelly, M. G., and Katona, S. (1966). An endogenous diurnal rhythm of bioluminescence in a natural population of dinoflagellates. *Biol. Bull.* **131**, 115–126.

Könitz, W. (1965). Elktronmikroskopische Untersuchungen an *Euglena gracilis* im tagesperiodischen Licht-Dunkel-Wechsel. *Planta* **66**, 345–373.

Leinweber, F. J. (1961). Temperature coefficient of endodiurnal leaf movements in *Phaseolus*. *Nature, Lond.*, **189**, 1028.

Lörcher, L. (1958). Die Wirkung verschiedener Lichtqualitäten auf die endogene Tagesrhythmik von *Phaseolus*. *Z. Bot.* **46**, 209–241.

MacDowall, F. D. H. (1964). Reversible effects of chemical treatments on the rhythmic exudation of sap by tobacco roots. *Can. J. Bot.* **42**, 115–122.

Moser, I. (1962). Phasenverschiebungen der endogenen Tagesrhythmik bie *Phaseolus* durch Temperatur- und Lichtintensitätänderung. *Planta* **58**, 199–219.

Oltmanns, O. (1960). Über den Einfluss der Temperatur auf die endogene Tagesrhythmik bei der Kurztagspflanze *Kalanchoë blossfeldiana*. *Planta* **54**, 233–264.

Overland, L. (1960). Endogenous rhythm in opening and odor of flowers of *Cestrum nocturnum*. *Am. J. Bot.* **47**, 378–382.

Palmer, J. D., and Round, F. E. (1965). Persistent, vertical-migration rhythms in benthic microflora. I. The effect of light and temperature on the rhythmic behavior of *Euglena obtusa*. *J. mar. biol. Ass. U.K.* **45**, 567–582.

Pirson, A. (1957). Induced periodicity of photosynthesis and respiration in *Hydrodictyon*. *In* Research in Photosynthesis. pp. 490–499 (ed. H. Gaffron *et al.*). Interscience Publishers, Inc., New York.

Pittendrigh, C. S. (1954). On temperature independence in the clock system controlling emergence time in *Drosophila*. *Proc. natn. Acad. Sci. U.S.A.* **40**, 1018–1029.

Pittendrigh, C. S., Bruce, V. G., Rosensweig, N. S., and Rubin, M. L. (1959). A biological clock in *Neurospora*. *Nature, Lond.*, **184**, 169–170.

Pohl, R. (1948). Tagesrhythmus im phototaktischen Verhalten der *Euglena gracilis*. *Z. Naturf.* **3b**, 367–378.

Richter G. (1963). Die Tagesperiodik der Photosynthese bei *Acetabularia* und ihre Abhängigkeit von Kernaktivität, RNS- und Protein-Synthese. *Z. Naturf.* **18b**, 1085–1089.

Rogers, L. A., and Greenbank, G. R. (1930). The intermittent growth of bacterial cultures. *J. Bact.* **19**, 181–190.

c

Round, F. E., and Palmer, J. D. (1966). Persistent, vertical-migration rhythms in benthic microflora. II. Field and laboratory studies on diatoms from the banks of the River Avon. *J. mar. biol. Ass. U.K.* **46,** 191–214.

Ruddat, M. (1961). Versuche zur Beeinflussung und Auslösung der endogenen Tagesrhythmik bei *Oedogonium cardiacum* Wittr. *Z. Bot.* **49,** 23–46.

Sargent, M. L. (1967). Biochemical, genetic, and physiological studies on a strain of *Neurospora crassa* exhibiting circadian conidiation. 141 pp. Ph.D. Thesis, Stanford University.

Sargent, M. L., and Woodward, D. O. (1967). Biochemical studies on an invertase-less strain of *Neurospora* exhibiting circadian conidiation. *Fedn Proc. Fedn Am. Socs., exp. Biol.* **726,** Abstract.

Schmidle, A. (1951). Die Tagesperiodizität der asexuellen Reproduktion von *Pilobolus sphaerosporus. Arch. Mikrobiol.* **16,** 80–100.

Schweiger, E., Wallraff, H. G., and Schweiger, H. G. (1964). Über tagesperiodische Schwankungen der Sauerstoffbilanz kernhaltiger und kernloser *Acetabularia mediterranea. Z. Naturf.* 9b, 499–505.

Semon, R. (1905). Über die Erblichkeit der Tagesperiode. *Biol. Zbl.* **25,** 241–252.

Sussman, A. S., Lowry, R. J., and Dukkee, T. (1964). Morphology and genetics of a periodic mutant of *Neurospora crassa. Am. J. Bot.* **51,** 243–252.

Sweeney, B. M. (1959). Endogenous diurnal rhythms in marine dinoflagellates Internat. Oceanographic Congress Preprints. pp. 204–207 (ed. M. Sears). AAAS, Washington, D.C.

Sweeney, B. M. (1960). The photosynthetic rhythm in single cells of *Gonyaulax polyedra. Cold Spring Harb. Symp. quant. Biol.* **25,** 145–148.

Sweeney, B. M. (1963). Biological clocks in plants. *A. Rev. Pl. Physiol.* **14,** 411–440.

Sweeney, B. M. (1963). Resetting the biological clock in *Gonyaulax* with ultra-violet light. *Pl. Physiol. Lancaster,* **38,** 704–708.

Sweeney, B. M. (1965). Rhythmicity in the biochemistry of photosynthesis in *Gonyaulax.* pp. 190–194 *in* Circadian Clocks (ed. J. Aschoff). North-Holland Publishing Co., Amsterdam.

Sweeney, B. M., and Hastings, J. W. (1958). Characteristics of the diurnal rhythm of luminescence in *Gonyaulax polyedra. J. cell. comp. Physiol.* **49,** 115–128.

Sweeney, B. M., and Hastings, J. W. (1957). Rhythmic cell division in populations of *Gonyaulax polyedra. J. Protozool.* **5,** 217–224.

Sweeney, B. M., and Hastings, J. W. (1960). Effects of temperature upon diurnal rhythms. *Cold Spring Harb. Symp. quant. Biol.* **25,** 87–104.

Sweeney, B. M., and Haxo, F. T. (1961). Persistence of a photosynthetic rhythm in enucleated *Acetabularia, Science, N.Y.,* **134,** 1361–1363.

Sweeney, B. M., Tuffli, C. F., and Rubin, R. H. (1967). The circadian rhythm in photosynthesis in *Acetabularia* in the presence of actinomycin D, puromycin, and chloramphenicol. *J. gen. Physiol.* **50,** 647–659.

Terborgh, J., and McLeod, G. C. (1968). The photosynthetic rhythm of *Acetabularia crenulata.* I. Continuous measurements of oxygen exchange in alternating light-dark regimes and in constant light of different intensities. (In press.)

Todt, D. (1962). Untersuchungen über Öffnung und Anthocyangehaltsveränder-ungen der Blüten von *Cichorium intybus* im Licht-Dunkel-Wechsel und unter konstanten Bedingungen. *Z. Bot.* **50,** 1–21.

Uebelmesser, E. R. (1954). Über den endonomen Tagesrhythmus der Sporangien-trägerbildung von *Pilobolus. Arch. Mikrobiol.* **20,** 1–33.

Vaadia, T. (1960). Autonomic diurnal fluctuations in rate of exudation and root pressure of decapitated Sunflower plants. *Pl. Physiol. Lancaster,* **13,** 701–717.

Vanden Driessche, T. (1966). Circadian rhythms in *Acetabularia*: photosynthetic capacity and chloroplast shape. *Expl. Cell Res.* **42,** 18–30.

Vanden Driessche, T. (1966). The role of the nucleus in the circadian rhythms of *Acetabularia mediterranea. Biochim. biophys. Acta* **126,** 456–470.

Wagner, R. (1963). Der Einfluss niedriger Temperatur auf die Phasenlage der endogen-tagesperiodischen Blattbewegungen von *Phaseolus multiflorus. Z. Bot.* **51,** 179–204.

Warren, D. M. (1964). Rhythmic carbon dioxide metabolism in plants. Ph.D. Thesis, University of London.

Warren, D. M., and Wilkins, M. B. (1961). An endogenous rhythm in the rate of dark fixation of carbon dioxide in leaves of *Bryophyllum fedtschenkoi. Nature Lond.* **191,** 686–688.

Wassermann, L. (1959). Die Auslösung endogen-tagesperiodischer Vorgänge bei Pflanzen durch einmalige Reize. *Planta* **53,** 647–669.

Wilkins, M. B. (1959.) An endogenous rhythm in the rate of carbon dioxide output of *Bryophyllum.* I. Some preliminary experiments. *J. exp. Bot.* **10,** 377–390.

Wilkins, M. B. (1960). An endogenous rhythm in the rate of carbon dioxide output of *Bryophyllum.* II. The effect of light and darkness on the phase and period of the rhythm. *J. exp. Bot.* **11,** 269–288.

Wilkins, M. B. (1960). The effect of light on plants. *Cold Spring Harb. Symp. quant Biol.* **25,** 115–120.

Wilkins, M. B. (1962). An endogenous rhythm in the rate of carbon dioxide output of *Bryophyllum.* III. The effects of temperature on the phase and the period of the rhythm. *Proc. R. Soc.* B **156,** 220–241.

Wilkins, M. B., and Holowinsky, A. W. (1965). The occurrence of an endogenous rhythm in a plant tissue culture. *Pl. Physiol. Lancaster,* **40,** 907–909.

Wilkins, M. B., and Warren, D. M. (1963). The influence of low partial pressures of oxygen on the rhythm in the growth rate of the Avena coleoptile. *Planta* **60,** 261–273.

Zimmer, R. (1962). Phasenverschiebung und andere Störlichtwirkungen auf die endogen tagesperiodischen Blütenblattbewegungen von *Kalanchoë bloss-feldiana. Planta* **58,** 283–300.

CHAPTER 4

Rhythms that Match Environmental Periodicities: Tidal, Semi-lunar and Lunar Cycles

Along the edges of the ocean, especially where there is a large tidal ebb and flow or a very gradual slope, the alternation of high and low tide is perhaps the most spectacular feature of the environment. Organisms which inhabit these regions are alternately exposed, then covered. The attendant changes in temperature and light intensity may be extreme. It would not be surprising, then, to find that organisms inhabiting the intertidal zone show tidal rhythms of some sort.

The clearest and best known tidal rhythms are those found in invertebrates from this region. One of the interesting examples of a tidal pattern comes from the experiments of Rao on the pumping rate of the mussels, *Mytilus californianus* and *M. edulis*. These bivalves pump more water through their gills at times when the tide is high than during low tide. When the amount of water passing through the animal is measured in the laboratory, the rate is found to vary, rising and falling in almost perfect synchrony with the water level at the site from which they were collected, even mirroring the variations in magnitude of tidal ebb and flow quite accurately. Furthermore, no differences in the course of this rhythm in pumping rate could be detected when the temperature was varied between 9° C and 20° C, or when the animals were kept in continuous darkness or continuous light. The same rhythm was found in mussels collected from a depth of 30 metres and those from 1 metre while animals which were growing on floats and so sensed no difference in water level still followed the same rhythmicity as those from nearby pier pilings. Such observations strongly suggested that the rhythm was not under the direct control of the tidal water level, but was originating from an endogenous cycle somehow synchronized to the tidal changes. That the cycles were indeed endogenous was further indicated by the observations that mussels collected on the east coast at Woods Hole and flown to California showed a tidal rhythm in the laboratory which continued in time with the tidal changes in their east coast home, out of synchrony with the west coast tidal changes. East coast mussels actually exposed to the tide hanging in baskets from the pilings of a pier in Corona del Mar, California, quickly reset their pumping rhythm to match west coast tides.

A number of invertebrates which live on beaches or tide flats have been observed to adjust their activities to the stage of the tide. The flat worm *Convoluta*, which contains green symbiotic algae, emerges from the sand only at low tide. On the other hand, the amphipod *Synchelidium* leaves the sand only at times when the water covers its beach habitat. Enright has shown that *Synchelidium*, brought into the laboratory and kept in DD in vessels containing sea water and sand, continues its cyclic swimming and resting. Like *Mytilus*, too, this creature mimics in the laboratory not only the tidal cycle in general but even the details of tidal fluctuations on its native beaches. The tidal changes in water level differ on different beaches and the shape of the tidal cycle is often considerably modified from a direct correspondence with the nightly position of the moon. While on most shores, there are two high and low tides a day, 12·4 hours apart, on others there is only a single tidal cycle. Both *Mytilus* and *Synchelidium* ape the local tidal changes so closely, that the tide itself rather than the moon seems to be the determining factor.

What evidence is there that animals can respond more directly to the apparent position of the moon in the sky? Papi has shown that another intertidal amphipod, *Talitrus*, seems to be able to navigate using the moon as a fix. These animals behave as if they were aware of the direction of the sea on their home beaches, since they take off in this direction when frightened, even when they have been transported to a location where the real position of the ocean is in the opposite direction. Their ability to orient with respect to the sea depends on their being able to see the moon at night. The simplest assumption is that they possess an endogenous rhythm with a period of 24·8 hours like the lunar day to which they compare the actual position of the moon so that they can correct for its progression across the sky. A circadian rhythm cannot be used for this purpose since the moon's period is longer than 24 hours. Not even an endogenous circadian rhythm with a free-running period longer than 24 hours can serve this purpose, since such a rhythm would be entrained by the daily light and darkness to an exact 24-hour period. Brown and his collaborators are able to detect components of a similar 'tidal' lunar-day period in their data for the colour changes, activity and respiratory oxygen uptake of crabs, and in fact for the metabolic activity of animals which are never ordinarily subjected to tidal variation like the earthworm and the rat. Taken together, these data offer strong support for the existence of endogenous tidal rhythmicity at least in some marine invertebrates. Are comparable rhythms also found in plants which inhabit intertidal regions?

For many years it has been noted that the mud of tidal flats changes colour shortly after it is exposed as the tide falls. The colour change is brought about by the accumulation at the surface of very large numbers of motile pigmented algae. Fauvel and Bohn (1907) were perhaps the first to investigate whether or not such an alga would continue to appear on the surface of mud samples brought to the laboratory and hence no longer under the direct influence of the tide. They found that the diatom *Pleurosigma* did alternately accumulate and disappear in mud samples for some time, and seemed to appear later each day, as did the tide.

Another example of this phenomenon was studied by Bracher in 1937. The mouth of the River Avon in Bristol, England, is an estuary with very large tidal differences in water level. The mud flats which are exposed at low tide soon become green in spots due to the accumulation of *Euglena obtusa* (called by Bracher *E. limosa*). She reported that when she brought the organisms to the laboratory, they continued to move up and down with a tidal period. However, recently Palmer and Round have reinvestigated the movements of this *Euglena* from the River Avon. They confirmed the field observation of Bracher, using small pieces of lens tissue laid on the mud to trap the organisms at the surface so that they could be counted, and by making miniature mud cores, freezing them and sectioning them in the frozen condition to determine where the organisms were found at different times in the tidal cycle.

Euglena cells did accumulate at the surface during low tides which occurred during daylight. Before the tide came in again, they moved down into the mud. None was found at the surface when low tide occurred at night. This would not be surprising if motion upward always consisted of swimming toward the light. Laboratory experiments brought to light an unexpected fact: as soon as this *Euglena* was brought into the laboratory and placed in LL 980 lux the cells displayed a typical circadian rhythm (Fig. 4-1). In constant light, the period was approximately the same at 5, 10, 15 and 18·5° C, the Q_{10} being about 0·9. The phase of this circadian rhythm was determined by the last dark period in nature, whether it were a night or a dark period resulting from the covering of the mud flat by murky water at high tide. Thus the rhythm was actually circadian, entrained by the tidally-caused light-dark cycle, rather than truly tidal. Round, Palmer and Happey also investigated a number of diatoms from the River Avon mud flats and these too were shown to possess circadian, rather than tidal endogenous rhythms. The question remained whether possibly all apparently tidal cycles in plants were merely manifestations of circadian rhythms reset by light intensity differences associated with the incoming tide.

Fauré-Fremiet meanwhile had made observations of the activities of the Chrysomonad *Chromulina psammobia* which also moves up at low tide to colour the mud flats of its habitat. He found that in this flagellate the direction of phototactic response reverses, being positive at low tide, negative at high tide. The presence of mud was not necessary

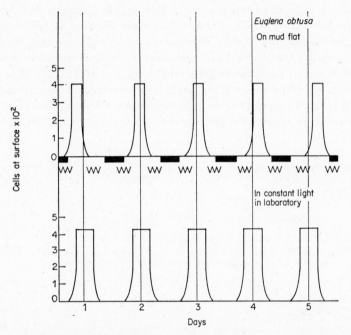

FIG. 4-1. The vertical migration rhythm in *Euglena obtusa* on a mud flat (top curve), and in constant light in the laboratory (bottom curve). The dark bars on the upper abscissa represent night, while the wavy lines show when the tide covered the mud flat. (Palmer and Round, 1965.)

for the observation of changes in phototaxis, which continued to occur in the laboratory for about 6–7 days after collection, more or less synchronized with the tide. The changes gradually became less marked until cells were always in the positively phototactic condition. In the next year, Fauré-Fremiet was visiting Woods Hole and while there noted that brown spots appeared at low tide on the mud flats of Barnstaple harbour. This time a diatom proved to be responsible, *Hantzschia amphioxys*. This diatom can glide half its length in a second. Cells examined in a drop of water were positively phototactic at low tide.

At high tide they moved away from light and seemed to exude a mucilaginous material which caused them to stick to each other and to sand grains. The rhythm disappeared after a few days in the laboratory.

After finding that the 'tidal' rhythm in *Euglena obtusa* and the diatoms of the River Avon was in reality a typical circadian rhythm reset by the murky water at high tide, Palmer and Round were interested to see whether this was also the case with *Hantzschia*, growing in clearer waters. Consequently, they reinvestigated the behaviour of *Hantzschia* collected from the same Barnstaple harbour and brought into the laboratory. Cells at the surface were counted as before by the lens tissue technique. Unlike *Euglena obtusa*, *Hantzschia* still showed a rhythm with a tidal period in the laboratory, migrating up at intervals of 24·8 hours in continuous light and even in LD 12 : 12, (FIG. 4-2). Circadian rhythms never show periods other than 24 hours when entrained by a 24 hour LD cycle. Another striking observation pointed to the presence in *Hantzschia* of a rhythm other than a modified circadian

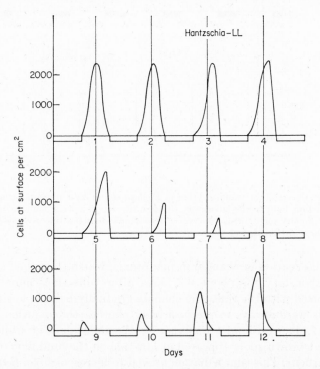

FIG. 4-2. The vertical-migration rhythm in the diatom *Hantzschia* during 12 days in the laboratory under constant illumination. The bars under the abscissa represent the night phase of a postulated circadian rhythm. (Palmer and Round, 1967.)

one. In LD, no migration took place during the dark period, and when cells were transferred to LL, still no migration ever occurred in what corresponded to the night phase of a circadian rhythm. The maximum in accumulation of cells at the mud surface took place later each day in accordance with the tidal cycle, until the night phase was reached. In the next cycle no migration was seen at all, and on the following day, a peak appeared in the morning. The interpretation offered by Palmer for this observation was that the true period of the tidal rhythm was 12·4 hours but the tidal maximum could not be expressed when it coincided with night phase of the circadian rhythm. Such a situation would result from the interaction of a diurnal rhythm in ability to move and a tidal rhythm in direction of movement. In any case a truly tidal rhythm seemed to be present in *Hantzschia*.

The evidence that there may be a tidal rhythm in *Fucus* comes from the work of Brown, Freeland and Ralph (1953). Indeed *Fucus* grows on rocks which are alternately exposed and covered due to tidal changes in water level, and one might expect to find it adapted to these environmental extremes. *Fucus* has considerable ability to resist desiccation, probably accomplished by the secretion of polysaccharide mucilage. Brown's data however, indicates that it also shows variations in respiratory rate in which both solar-day and lunar-day components can be recognized. The variations are rather small however under laboratory conditions, and the use of such devices as sliding averages are necessary to demonstrate their existence.

A tidal component has been postulated for the photosynthesis of *Dictyota*. Oxygen evolution from the thallus of this brown alga varies during the course of a day, but the time of the apparent maximum changes, being later every day, in accord with a 24·6 hour period. The variation appears to be due to the interaction of a tidal with a diurnal rhythm, in which the tidal effect is inhibitory, taking a progressively later bite out of the diurnal maximum each day, as if a shadow of the moon were eclipsing the daily photosynthesis. The importance of this tidal rhythm will be discussed below as part of the interpretation offered by Müller and Vielhaben for the semilunar rhythm in reproduction which they demonstrated in *Dictyota*.

The tides are caused by the earth's rotation and the moon's movement in its orbit around the earth. Still another periodicity results from the relative motion of earth, moon, and sun. A new moon, then a full moon appear once in each lunar month of 29·5 days. Twice each lunar month, or every 14·8 days, the tidal changes in water level are more extreme than usual. This occurs at new and full moon when the sun and the moon are pulling along the same line, and create a higher tide, the 'spring' tide. When the moon is in its first or third quarter and the sun and moon pull at right angles to each other, the tidal range is least. Such tides are known as 'neap' tides. Are there any rhythmicities in plants or animals which match these lunar and semilunar cycles?

Very little is known concerning even the existence of rhythmicities with periods which match either the lunar month or the semilunar spring tides. There are, however, among marine animals, a few examples which make up for their rarity by their spectacular nature. Perhaps the most famous example concerns the behaviour of the grunion, a small fish which inhabits the sea off California. This fish spawns on the nights just after the highest spring tide twice each lunar month in summer with quite remarkable predictability.

Even more surprising is the behaviour of this lithe little fish, which actually leaves the water and wriggles up the beach to lay eggs at the level reached by the highest tide or to fertilize the eggs as they are discharged. The fish may wriggle up the beach 5 yards or more and return to the sea on the next wave. This behaviour cannot so far be studied under controlled conditions so that it is not known whether an endogenous cycle cues the grunion concerning when it is time to spawn.

The reproductive part of the Palolo worm (*Eunice vividis* and *E. fucala*) comes to the ocean surface to spawn only during a few nights at the last quarter of the moon. *Platynereis* also regulates its reproductive cycles by the phase of the moon. This marine polychaet has been studied under laboratory conditions by Hauenschild. In this case, the animals respond directly to moonlight. Metamorphosis into the sexually capable form is followed by swarming of the reproductive parts which separate from the rest of the animal. In the laboratory these events continue, but if the animals are kept on an artificial LD 12 : 12, spawning is distributed in time in a random manner. However, if 6 days of LL are interposed in the LD schedule, this 'night light' serves to synchronize metamorphosis, the end of the LL period being the timing cue. Synchrony in spawning is soon lost however, and there is no evidence for a rhythmicity of a lunar type.

Field observations have shown that a number of marine algae liberate either gametes or spores at the time of the spring tides, twice every lunar month. In the laboratory such long rhythms are difficult to follow. However, at least two algae have been studied under conditions where light and temperature were controlled, *Dictyota* and *Halicystis*.

Müller was able to grow *Dictyota dichotoma* in culture and obtain fruiting which he measured by counting the number of eggs released. When pieces of the thallus of this brown alga were cultivated under natural light, gametes were produced in bursts which occurred every

14–15 days. However, when they were moved to an artificial LD cycle of 14 : 10 LD, few eggs were released and no synchrony could be seen. An explanation for these apparently contradictory findings was suggested by the experiments of Hauenschild: perhaps as in *Platynereis*, light at night, such as moonlight under natural conditions for example, may be necessary to synchronize the production of gametes. Müller tested this hypothesis by leaving the lights on during one night, thus exposing his *Dictyota* to a simulated moon. Ten days later a burst of

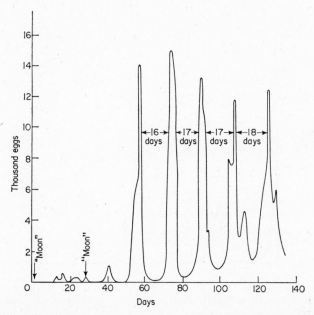

FIG. 4-3. Egg production by the brown alga *Dictyota dichotoma* in the laboratory growing with a 14 : 10 LD cycle, 20° C. The dark period was replaced by light on day 1 and day 28, as shown by the arrows on the abscissa, to give the plants artificial 'moonlight'. (Müller, 1962.)

egg production was indeed found and unlike the *Platynereis*, a cycle in reproduction continued to be expressed in which maxima occurred every 16 days thereafter, for at least 5 semilunar cycles (FIG. 4-3). At night, an intensity of 3 lux was enough to produce this effect. To account for the semilunar nature of the rhythm observed, Müller suggested that both a tidal and a diurnal rhythm was present in *Dictyota* and that these two periodicities interacted. When a period of 24 and a period of 12·4 hours are present, simultaneously affecting the same physiological function, the maxima of these two periodicities will coincide at intervals of about 15 days. If the resulting reinforcement or 'beat' could conceivably cue a response, then this response will show a period of about 15 days, a semilunar periodicity. This hypothesis could be tested by changing the frequency of one of these rhythms and hence changing the position of the beat. The way in which tidal rhythms can be reset in plants is unknown, but circadian rhythms can easily be entrained to a different period by LD cycles of the desired period. Müller began a series of experiments in which *Dictyota* was maintained on a 23·5 hour light-dark regime, a schedule which theoretically should result in a

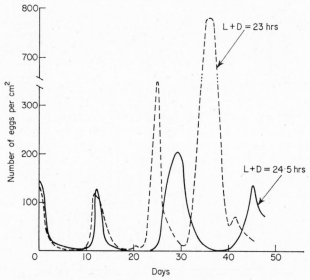

FIG. 4-4. The production of eggs by *Dictyota dichotoma* growing in the laboratory with light-dark cycles shorter than 24 hours (13·5 : 9·5 = 23) (dashed line); and on light-dark cycles longer than 24 hours (14·25 : 10·25 = 24·5 hours) (solid line). (Vielhaben, 1963.)

beat every 12 days instead of every 15. He did indeed find a shortening of the next interval following the initiation of such an LD, but only to 13–14 days. His results were extended by Vielhaben who examined fruiting of *Dictyota* on 23 and 24·5 hour cycles of LD (FIG. 4-4). Shortening the LD cycle did appear to shorten the period of the reproduction, L : D 13·5 : 9·5 giving reproduction at 11–12 day intervals while LD 14·25 : 10·25 lengthened the period of egg release to 16–17 days. Vielhaben's controls were apparently reproducing at 14–15- rather than at 16-day intervals as had Müller's, but no data for controls run at the same time as experimental schedules are given in her paper. Calculation shows that if the LD schedule is shortened to a 23 hour cycle, then beats should occur after about 12 days, as she observed, but lengthening the LD cycle to 24·5 hours would be expected not to produce a beat for 55 days, rather than the observed 16–17. Since light during the night part of the LD cycle induces fruiting, one must consider whether this light really may have coincided with the night phase of the lunar rhythm. If this were the condition necessary for induction of fruiting, then changing the LD schedule could have resulted in light falling during a lunar night phase somewhere in the altered schedules. This assumption requires a 24·8 hour lunar day rhythm rather than a 12·4 hour tidal one. Actually there is no direct evidence in *Dictyota* for a *tidal* 12·4 hour rhythm. The evidence for a component with a 24·8 hour period is the shadow cast across the photosynthesis curve referred to previously, which begins 0·8 hours later each day. A diurnal rhythm in the actual time at which gametes were released which continued in LL was demonstrated by Vielhaben in *Dictyota*.

In 1936, Hollenberg reported that *Halicystis*, the gametophytic stage of the coenocytic green alga *Derbesia*, which had been observed to form gametangia only at especially low 'spring' tides twice each lunar month, could carry on this periodicity in the laboratory, removed from tidal influences. The exact conditions under which he cultured his algae are unclear, however. Recently the reproduction of *Halicystis parvula* the gametophyte of *Derbesia tenuissima*, has been investigated under carefully controlled conditions by Ziegler Page. She finds that this *Halicystis* does not show a semilunar periodicity under laboratory conditions, either in LD 12: 12 or in LL. Under both these conditions, reproduction occurs quite regularly and in synchrony, however. The interval between successive gametangia is usually 4–5 days when growth conditions are optimum. This is a far cry from a semilunar period. She repeated the experiments of Vielhaben, using *Halicystis*. However, no effect of light at night on the time of production of gametangia could be observed. Furthermore, keeping the algae in LD 8 : 8, a 16-hour cycle, had no effect on the time of initiation of gametangia (FIG. 4-5). *Halicystis* was shown by Ziegler Page and Sweeney to possess a well-defined circadian rhythm in photosynthesis persisting in LL. There was no evidence of any effect of a beat between this diurnal and a tidal rhythm, however, nor was there any shadow cast by a tidal rhythm on the daily photosynthesis, as in *Dictyota*, the two plants showing completely different behaviour.

When *Halicystis* was maintained under lower than optimal light intensities, reproduction still occurred. The most interesting finding from these experiments was that under these conditions, the interval between the formation of successive gametangia on the same plant was

either twice or three times the normal 4–5 days, but intermediate intervals, such as one and a half, or two and a half times normal, were not
observed (FIG. 4-6), as if a rhythm in gametangial formation with a
4–5 day period were still present, and reproduction could not take place
except at a certain phase of this rhythm. When conditions were unfavourable for the formation of gametangia, some necessary prerequisite
was not present when the first phase allowing reproduction occurred,
and plants were forced to await the coming of a second, or perhaps even
a third, permissive phase to initiate gametangia. Such evidence as this

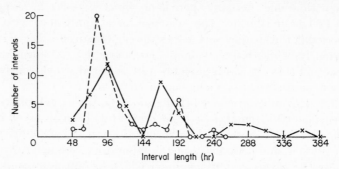

FIG. 4-5. The interval between successive gametangia in *Derbesia tenuissima* growing
on LD 8 : 8, compared to that when plants are grown on LD 12 : 12. Changing the light-
dark schedule did not drastically alter the timing of gametangial formation. (Ziegler
Page and Sweeney, 1968.)

from the distribution of intervals between successive gametangia
provides the strongest evidence that a rhythm with a period of 4–5 days
is really present and controls gametangial formation.

The field observations that *Halicystis* reproduces with a semilunar
periodicity remain to be explained. It is possible that in nature, gametangia are formed on only every fourth cycle of the endogenous rhythm.
On the other hand, a direct effect of the spring tide may be involved in
initiating gametangia. Ziegler Page and Sweeney found that if vegetative plants were exposed to an increase in temperature, even a rather
small and transitory one, all reproduced three days later. Perhaps such
an increase might be associated with spring tides. In any case, more
careful observations of *Halicystis* in nature are desirable.

The control of environmental conditions for the length of time necessary for the study of such long period rhythms proves difficult in
practice. However, precise knowledge of semilunar and lunar rhythms
will only be gained through such experiments.

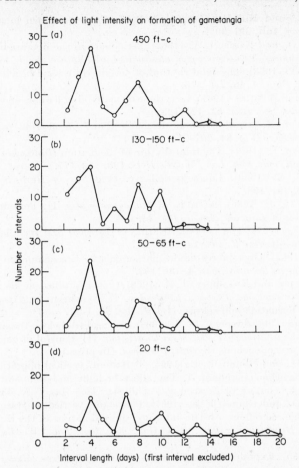

FIG. 4-6. The intervals between successive gametangia in *Derbesia tenuissima* grown at different light intensities. Note that there are maxima at about 4-, 8- and 12-day intervals, but not at 5- and 6-day intervals. (Ziegler Page and Sweeney, 1968.)

REFERENCES

Bracher, R. (1937). The light relations of *Euglena limosa* Gard. Part I. The influence of intensity and quality of light on phototaxy. *J. Linn. Soc. (Bot.)* **51**, 23–42.

Brown, F. A., Jr., Freeland, R. O., and Ralph, C. L. (1955). Persistent rhythms of O_2 consumption in potatoes, carrots and the sea weed, *Fucus*. *Pl. Physiol. Lancaster*, **30**, 280–292.

Brown, F. A. Jr. (1957). Biological Chronometry. *Am. Nat.* **91**, 129–133.

Brown, F. A. Jr. (1959). Living clocks. *Science N.Y.* **130**, 1535–1544.

Bünning, E., and Müller, D. (1961). Wie messen Organismen lunare Zyklen? Z. Naturf. **16b**, 391–395.

Christie, A. O., and Evans, L. V. (1962). Periodicity in the liberation of gametes and zoospores of *Enteromorpha intestinalis* Link. *Nature, Lond.*, **193**, 193–194.

Enright, J. T. (1963). The tidal rhythm of activity of a sand-beach Amphipod. *Z. vergl. Physiol.* **46**, 276–313.

Fauvel, P., and Bohn, G. (1907). Le rythme de marée chez les diatomées littorales. *C.r. Séanc. Soc. Biol.* **62**, 121–123.

Fauré-Fremiet, E. (1950). Rythme de marée d'une *Chromulina* psammophile. *Bull. biol. Fr. Belg.* **84**, 207–214.

Fauré-Fremiet, E. (1951). The tidal rhythm of the diatom *Hantzschia amphioxys*. *Biol. Bull. mar. biol. Lab., Wood's Hole*, **100**, 173–177.

Hauenschild, C. (1960). Lunar periodicity. *Cold Spring Harb. Symp. quant. Biol.* **25**, 491–498.

Hollenberg, G. J. (1936). A study of *Halicystis ovalis*. II. Periodicity in the formation of gametes. *Am. J. Bot.* **23**, 1–3.

Hoyt, W. D. (1927). The periodic fruiting of *Dictyota* and its relation to the environment. *Am. J. Bot.* **14**, 592–619.

Müller, D. (1962). Über jahres- und lunarperiodische Erscheinungen bei einigen Braunalgen. *Botanica mar.* **4**, 140–155.

Page, J. Ziegler, and Kingsbury, J. M. (1968). Culture studies on the marine alga *Halicystis parvula—Derbesia tenuissima*. II. Synchrony and periodicity in gamete formation and release. *Am. J. Bot.* **55**, 1–11.

Page, J. Ziegler, and Sweeney, B. M. (1968). Cultural studies on the marine green alga *Halicystis parvula—Derbesia tenuissima*. III. Control of gamete formation by an endogenous rhythm. *J. Phycol.* (In press.)

Palmer, J. D., and Round, F. E. (1965). Persistent, vertical-migration rhythms in the benthic microflora. I. The effect of light and temperature on the rhythmic behavior of *Euglena obtusa*. *J. mar. biol. Ass. U.K.* **45**, 567–582.

Palmer, J. D., and Round, F. E. (1967). Persistent, vertical-migration rhythms in benthic microflora. VI. The tidal and diurnal nature of the rhythm in the diatom *Hantzschia virgata*. *Biol. Bull. mar. biol. Lab., Wood's Hole*, **132**, 44–55.

Papi, F. (1955). Experiments on the sense of time in *Talitrus saltator* (Montagu) (Crustacea-Amphipoda). *Experimentia* **11**, 201–202.

Rao, K. P. (1954). Tidal rhythmicity of rate of water propulsion in Mytilus and its modifiability by transplantation. *Biol. Bull. mar. biol. Lab., Wood's Hole*, **106**, 353–359.

Round, F. E., and Happey, C. M. (1965). Persistent, vertical-migration rhythms in benthic microflora. IV. A diurnal rhythm of the epipelic diatom association in non-tidal flowing water. *Bull. Br. psychol. soc.* **2**, 463–471.

Round, F. E., and Palmer, J. D. (1966). Persistent, vertical-migration rhythms in benthic microflora. II. Field and laboratory studies on diatoms from the banks of the River Avon. *J. mar. biol. Ass. U.K.* **46**, 191–214.

Smith, G. M. (1947). On the reproduction of some Pacific coast species of *Ulva*. *Am. J. Bot.* **34**, 80–87.

Vielhaben, V. (1963). Zur Deutung des semilunaren Fortpflanzungszyklus von *Dictyota dichotoma* Z. f. Bot. **51**, 156–173.

Ziegler, J. R., and Kingsbury, J. M. (1964). Cultural studies on the marine alga *Halicystis parvula—Derbesia tenuissima*. I. Normal and abnormal sexual and asexual reproduction. *J. Phycol.* **4**, 105–116.

CHAPTER 5

Rhythms that Match Environmental Periodicities: The Year

In all but the equatorial regions of the world there is a change in the environment with season, a yearly cycle in temperature, light and sometimes rainfall. These seasonal changes may be severe in temperate latitudes. Plants and animals have evolved a whole series of adaptive changes which allow them to survive through the inclement part of the year. Plants form seeds or tubers, while vertebrates hibernate and insects go into diapause. On the other hand, growth and reproduction take place during that part of the year when mild weather will favour the survival of offspring.

The key to successful adaptation to a recurrent pattern of change is preparedness. Thus, as one would expect, it is not the coming of the conditions for which the adaptations are designed that sets off their initiation. A seed that isn't formed until the snow flies might never be formed at all. The question that interests us here is what is the nature of the devices which make possible the detection of the coming of winter or the coming of spring before either actually arrives. There are at least two possibilities which come to mind. First, organisms might contain mechanisms which are able to measure time in years, a rhythm with an annual period, just as circadian rhythms may be used to measure days. Secondly, organisms might make use of the fact that the day length changes with the seasons, becoming shorter as winter approaches and longer with the coming of spring. They would then only require a device for measuring the length of the day with respect to the night. It seems reasonable that one might find both of these methods in use, were one to canvass enough different organisms in one's search. It has been found, however, that the second method of predicting the coming season, the measurement of the length of the day, appears to be far the most common, both in plants and animals. I shall discuss this kind of yearly periodicity, photoperiodicity, first and then return to a consideration of the evidence for the existence of true annual rhythms.

The Measurement of Day Length in Photoperiodism

The importance of the length of day in initiating the change from vegetative to reproductive phases of growth was first discovered by Garner and Allard (1920), in tobacco and soy bean, plants which they showed initiated flower buds only when the day was *shorter than* a certain number of hours. Other plants were found to flower only when the days were *longer than* a given duration. The critical number of hours in both short-day and long-day plants was about but not necessarily exactly 12 hours, and varied from species to species. Both over-wintering conditions and the anatomical and behavioural preparation for reproduction in animals was later found to be cued by day length in much the same way as in plants. The question of how these organisms are able to measure the day or the night length in order to respond to a short or a long day became very interesting.

Two kinds of theories evolved to account for this time measurement and the proponents of each defended their views with vigour if not rancour. Just as time may be measured by using an hourglass or an electric clock, so could the length of day be sensed by the accumulation or disappearance of some cellular component (the sand of the hourglass), or by a comparison of the external light conditions with an internal oscillation with a period of 24 hours, a circadian clock in every way like those manifest in circadian rhythms of various kinds.

Proponents of the 'hourglass' theories for the time measurement in photoperiodism hunted for the component of the photoperiodic process which appeared or disappeared in light or darkness at a rate that would provide the necessary yardstick. It was soon apparent that in short-day plants at least, the length of the night rather than the day length was the parameter being sensed. This conclusion was reached as a result of experiments in which the night was interrupted by a very brief light exposure, only a few minutes. In short-day plants, given a short-day long-night regime, such an interruption completely blocked flowering, although the control plants given the same LD schedule but without the light interruption flowered as usual. A similar situation was discovered in long-day plants as well, where however, as expected, the long night prevented flower formation in the controls, and the light interruption now invoked flowering. Thus, short days and long nights interrupted by a very short light exposure were comparable in every way to long days and short nights.

Action spectrum determinations for the effect of light as an interruption in a long night (Borthwick, Hendricks, Parker, Toole and Toole, 1952) showed that this light effect was mediated by phytochrome in the far-red absorbing form, P_{fr}. Furthermore phytochrome could be switched from one form to another by light. Therefore what more natural than to suggest that the substance timing the length of the night was phytochrome? It was suggested (Hendricks, 1960, 1963) that the disappearance of P_{fr} and the appearance of P_r constituted the hourglass sand. It remained however to show that these two forms of phytochrome actually shifted in concentration in the manner required.

It has been found that P_{fr} is the favoured form of phytochrome at the end of a long exposure to white light such as a day. Thus the night begins with most of the phytochrome in this form. In a number of experiments, the state of phytochrome in etiolated seedlings during a long dark period has been studied. The seedlings were first irradiated briefly with red light with the intent of thus converting all the phytochrome present to P_{fr} at the beginning of the experiment. The amount of photoreversibility on irradiating with red after various lengths of time in the dark was determined and used as a measure of the reappearance of the red-absorbing form of phytochrome (P_r). Many tissues studied in this way seemed to show a disappearance of P_{fr} and a reappearance of P_r as darkness continued. This conversion was not rapid and so was invoked as the timer for the measurement of the night length. Even at first sight however, the time of conversion provided a serious difficulty for such a theory. In cotyledons of *Zea mays* and other monocotyledonous plants, changes in the form of phytochrome appeared to be completed after 4 hours of darkness (Butler *et al.* 1963). In etiolated seedlings of dicots such as *Pisum* and *Sinapsis* in darkness, the bulk of the P_{fr} disappeared even more quickly. Therefore why should the critical night lengths be not 4 but about 12 hours? Furthermore it has been shown (Salisbury, 1963) that the critical night length does not vary much as the temperature is changed. The conversion of phytochrome shows a much greater temperature dependence.

Another observation that does not fit the phytochrome hourglass model concerns the effect of cobaltous ion. In peas, Furuya *et al.* (1964) have shown that the presence of cobaltous ion (1×10^{-5}M) does not alter the changes of phytochrome in the dark. However, this ion increases the critical night length significantly in the short day plant *Xanthium* (Salisbury, 1959). Possibly, however, the measurement of changes in phytochrome in plants such as pea, which flower without regard for day length, does not provide an accurate picture of the situation in plants where flowering is under photoperiodic control. To test this possibility, Hopkins and Hillman examined the dark changes in the phytochrome in long-day, short-day and day-neutral monocot and dicot species. Coleoptiles of the long-day plant *Hordeum* and two strains of *Zea mays*, one a short-day plant and one day-neutral, all behaved in a very similar fashion with regard to the disappearance of P_{fr}, which was complete in 4–5 hours. The hypocotyls of the dicots examined also behaved similarly, whether they were long-day plants (*Raphinus sativus, Phaseolus coccineus*), day-neutral (*Phaseolus vulgaris*), or short-day plants (*Glycine max* and *Xanthium*) as adults. No disappearance was noted beyond 4 hours in darkness.

Recently the experiments of Hillman and others have shown that the apparent conversion of P_{fr} to P_r in darkness is at least in part an artifact. The misconception concerning dark conversion arose from the assumption that all the phytochrome had been transformed to P_{fr} by the exposure to red light before the beginning of the dark period. This was shown not to be the case, a certain portion always remaining in the P_r form no matter what the irradiation, on account of the overlapping of the absorption spectra of the two forms of phytochrome. Since the P_r remaining after red irradiation is not reversible, it cannot be detected at once after darkening the plant, but gradually becomes partially reversible as a portion of the P_{fr} disappears via other transformations in darkness. When a correction for this error was applied to the data, it became apparent that there was no dark reversion from P_{fr} to P_r in monocots. In dicots, the dark conversion, though probably real, was small compared to the disappearance of P_{fr} by other routes in the dark. Thus an hourglass composed of the reversion of P_{fr} to P_r seemed unlikely. Kinetics of the disappearance of P_{fr} do not encourage the view that this process could be responsible for photoperiodic timing either.

Moreover, all measurements of *in vivo* phytochrome must be carried out with etiolated tissue since the absorption of chlorophyll in the red part of the spectrum offers serious interference. Thus it is not possible to make measurements of the changes in phytochrome in the kind of plant tissue which is capable of the measurement of night length, namely green leaves. The rates of conversion or disappearance in such material could of course be quite different from that in etiolated seedlings.

What then of the endogenous circadian clock as a timer for the measurement of the length of the night? Bünning proposed in 1936 that the same timing mechanism was responsible for photoperiodic timing and the kind of diurnal rhythm that he was studying in bean leaf sleep movements. He distinguished two alternating phases of such cycles which he called 'photophil' or light-loving and 'scotophil' dark-loving. Photophil is equivaent to the day phase and scotophil the night phase of a circadian rhythm. He postulated that light falling on the plant in the scotophil phase would inhibit flowering; in the photophil enhance it.

The change from one state to the other would occur about every 12 hours hence accounting for the length of the critical day in both short and long-day plants. While short-day plants fit into this scheme further assumptions are needed to account for the existence of long-day plants. Bünning postulated that in these plants the photophil lasts longer than 12 hours thus extending into the first part of a night except during the long days of summer. If the circadian rhythm were really responsible for measurement of the day length, then leaf position should indicate whether or not a plant with a sleep-movement rhythm was in the photophil or the scotophil phase at any given time. Bünning achieved some success in applying this criterion, especially with regard to *Madia elegans*, the leaves of which spread open and close twice instead of the usual once a day. This plant was subsequently shown to flower on either quite short or very long days but not on days of intermediate length. That all rhythms should show maxima or minima coinciding with the middle of the photophil or scotophil soon proved to be an oversimplification however. The position of the maximum of one rhythm need not be the same as another in the same plant, as shown in *Gonyaulax* and many other systems.

Bünning's theory was at first considered rather bizarre by some scientists who were unfamiliar with circadian rhythms. However, support was shortly forthcoming from an unexpected quarter. Claes and Lang (1947) discovered that if the long-day plant *Hyoscyamus* was grown with nights 48 hours long, flowering was promoted as expected by light interruptions early in this dark period. The effectiveness increased to a maximum as the interruption was placed later and later, then, as still later interruptions were assayed, the effect decreased. Still later interruptions, however, again brought about the initiation of flowers. As an explanation for the unexpected finding that there were two peaks in promotion of flowering, Claes and Lang postulated that a light break that occurred in the early part of the night simply served as an extension of the previous light period, while those breaks in the latter part of the night were essentially added to the day following. Thus a short day preceded or followed by a light break was interpreted by the plant as a long day. However, it could equally well be a demonstration of a rhythm in some process involved in the light effect on flower initiation. This last interpretation was favoured by the discovery that, in the short-day plant *Kalanchoë*, light interruptions in an even longer night (Carr, 1952) also showed cycles of effectiveness, this time with three maxima about 24 hours apart. However, these peaks were of somewhat unequal amplitude, the middle and crucial peak sometimes being disappointingly indistinct.

Cycles in the effectiveness of light interruptions were discovered in a number of other short-day plants, and their presence in the long-day plant *Hyoscyamus* was confirmed (Clauss and Rau, 1936), although two other long-day plants studied by Hussey, *Anagallis* and *Arabidopsis*, failed to show rhythmicity.

Short-day plants proved especially well suited to the investigation of light interruptions in long nights. Among such experiments, Engelmann's with *Kalanchoë* are interesting in two respects. Remember that the flowers of this plant open and close in accord with a circadian rhythm in darkness. The position of the petals can thus be used to

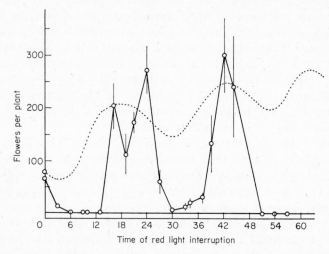

Fig. 5-1. The number of flowers induced in *Kalanchoë* by a 62-hour dark period interrupted with red light at different times. Lines through the points represent uncertainty. The dotted curve is the course of petal movement during a similar long dark period. (After Engelmann, 1960.)

follow the rhythm during a long dark period, although of course flowers are not available on the very plants where induction of flowering is being attempted. Secondly, it is possible to give this plant five short days (LD, 9:15) without inducing flowering. These five cycles of light and darkness prime the flowering process, so that only one more long dark period is necessary. When the length of this sixth dark period is varied, a marked rhythmicity can be observed, promotion in general being correlated with treatments where the end of the dark period coincides with the day phase behaviour of the flowers under similar conditions (FIG. 5-1).

Between 1957 and 1963, a long series of experiments were undertaken by Hamner and his collaborators Blaney, Nanda, Coulter and

Carpenter, in which the effect of varying the length of the dark period on flowering of the soya bean *Glycine max* was investigated. The light period was usually 8 hours long, well below the critical day length for soya bean. Light-dark cycles as long as 72 hours were studied. The results showed that there was a distinct rhythmicity in production of flowers with maxima about 24 hours apart, interspersed with cycles that inhibited flowering. The results of these experiments agree well with

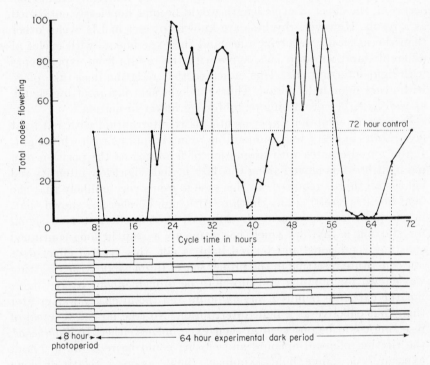

FIG. 5-2. Flowering of soya bean following 7 cycles of 8 hours light, 64 hours of darkness interrupted at various times with a 4-hour interruption with bright white light. The abscissa is the time in the dark period when the light interruption began. The light schedule is diagramed below. (After Coulter and Hamner, 1964.)

those on light interruptions, since the light coming on at the end of the dark period acted like a light break, inhibiting when it fell in the scotophil and promoting in the photophil phase of the postulated circadian rhythm (FIG. 5-2). In these experiments a second effect of light must be invoked however. In soy beans, at least seven cycles of light and darkness of the proper length are necessary for flower initiation. In general, in fact, more than one day of the required number of hours is necessary for induction of flowering in most plants. Hence cycles must be repeated. If the light and darkness of the artificial cycle sum to a multiple of 24 hours, as is the case in the experiments with light interruptions, each time the light period begins, it starts at the same phase of the circadian rhythm. Thus, the repetition of cycles does not lead to complicated phase relationships. However, in light-dark schedules the sum of which is not 24 hours, each cycle will begin at a different phase of the endogenous rhythm and unless resetting has occurred, the effect of cycle length would become hopelessly entangled as a result. However, rhythms are known to reset in LD cycles other than the naturally occurring ones and the experiments with cycles of different duration do in fact, agree with the results from experiments with light interruptions. This indicated that resetting does take place with every main light period. Thus, the light acts as a resetting signal, as well as either promoting or inhibiting flower formation.

In light interruption experiments, short interruption with red light give results in every way comparable to interruptions with white light. Inhibition alternates with promotion of flowering as the position of a red light break is moved along the dark period. Short low-intensity red light is all that is required. In this case it seems very unlikely that the light break is resetting any rhythm. Although Lörcher has shown that, in the bean leaf movement rhythm, red light is effective as a signal for starting a rhythm which has come to a stop in long-continued constant conditions, and far-red light is not, this type of effect does not constitute resetting in the proper sense. In *Phaseolus*, true resetting was shown by Bünning and Moser to be brought about by red light, but the exposure required was $3\frac{1}{2}$ hours and the intensity was greater than that effective as a light interruption. No true far-red reversal of resetting could be observed in their experiments either. In *Kalanchoë*, the rhythm of petal opening and closing could be reset by a 2-hour exposure to orange light (Zimmer, 1962), again a relatively long exposure. The lack of resetting by a 5-minute exposure to red light is nicely demonstrated in experiments with *Pharbitis nil*, in which the rhythm of effectiveness of an exposure to red light was unaltered by a second preliminary exposure (FIG. 5-3).

Uncertainties arising from the possibility of resetting could be avoided, were it possible to induce flowering with a single dark period of the proper length. This is fortunately possible with several plants. However, when such plants are induced under optimal conditions, rhythms are difficult to detect by the long night method, since the die is cast for flowering during the first 12 or so hours of darkness and cannot be reversed subsequently. Seedlings of a selection of *Cheno-*

podium rubrum discovered and studied by Cumming *et al.* do not show
this disadvantage. While induction can be achieved by a single night,
quantitative differences in effectiveness during a long dark period can
still be seen. Light interruptions and changes in the night length yield

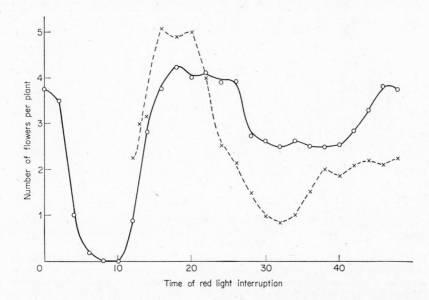

FIG. 5-3. Flowering of *Pharbitis nil* given a single 48-hour dark period which was interrupted by either a single 5-minute red light exposure at different times ○, or two such
red light exposures, the first at hour 12 of the dark period, the second at different times
as shown on the abscissa, X. Temperature 18 or 18·5°. Before the dark period, the plants
were in continuous light. (Redrawn from Takimoto and Hamner, 1965a and 1965b.)

quite comparable results (FIG. 5-4), and both clearly show a rhythm
in the effectiveness of light with a period of about 24 hours, continuing
for 2–3 cycles. Recently, it has been possible to extend these cycles by
supplying the plant with sugar (Cumming, 1967).

If conditions are made somewhat less than optimal for flowering,
rhythmic phenomena can be uncovered in two other short-day plants
which flower with a single inductive cycle, *Xanthium pennsylvanicum*
and *Pharbitis nil*. Salisbury (1965) showed that if *Xanthium* plants from
constant light were exposed to one very short night (7·5 hours), then
to a 'day' of only a few minutes followed by another dark period, a
rhythm in the effectiveness of this second dark period which depended
on its length could be observed (FIG. 5-5), maxima in flowering occurring when this second night was 24 or 48 hours long. Salisbury also
succeeded in showing that, for *Xanthium*, the light period was most

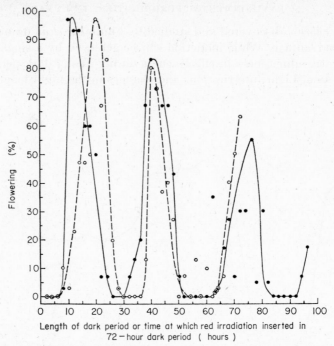

FIG. 5-4. The flowering of *Chenopodium rubrum* following a single dark period of different lengths from 3 to 96 hours ●, or a single 72-hour dark period interrupted at various times by 4 minutes of red light, ○. Plants were otherwise in continuous light. (After Cumming, Hendricks and Borthwick, 1965.)

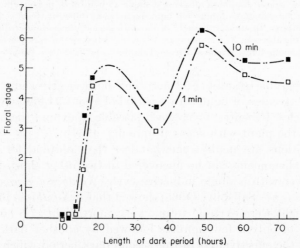

FIG. 5-5. The induction of flowers in *Xanthium pennsylvanicum* following treatment with two dark periods separated by either 1 (□) or 10 minutes (■) of light. The first dark period was 7·5 hours long in all cases, while the length of the second dark period was varied as shown on the abscissa. (After Salisbury, 1965).

effective when it also fitted into the framework of the circadian cycle and was about 12 hours long (FIG. 5-6).

In *Pharbitis*, it was necessary to resort to cold temperatures to show rhythmicity in the effectiveness of light interruption. At 20° C and 25° C, there is just a single maximum in inhibition of flowering by light breaks, 8 hours after the beginning of the dark period, since a single

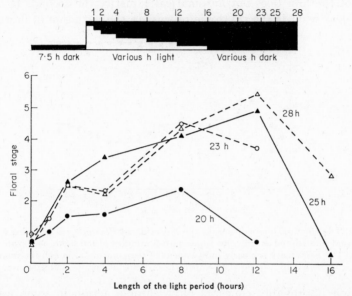

FIG. 5-6. The flowering of *Xanthium pennsylvanicum* following light-dark schedules consisting of a 7·5-hour dark period, then a light period and a subsequent dark period, each of variable length, as shown in the diagram above the curves. The total number of hours for the light and the second dark period is written beside the curve to which it applies. (After Papenfuss and Salisbury, 1966.)

long night induces flowering at these temperatures. However, if the temperature is reduced to 18·5° C or lower, then two maxima in inhibition can be seen clearly (FIG. 5-7). This observation supports the rhythm hypothesis, since it is now well substantiated in many rhythmic systems that circadian periods are very little affected by the temperature level.

The discovery of what appears to be a *bona fide* circadian rhythm in the response of long- and short-day plants to light, a rhythm which continues for at least three cycles in continuous darkness, poses a most serious difficulty to any hourglass theory, since timers of the hourglass type can only measure one time interval, unless they are turned over. Bünning's hypothesis is thus favoured by the experiments on flower induction in cycles where the dark period is much longer than the critical night.

Inherent in Bünning's explanation for the measurement of day length in photoperiodic organisms are two postulates, first, that there is a circadian rhythm in some function vitally associated with flower induction in both short- and long-day plants; and second, that the organism makes use of this rhythm for time measurement. The first assertion now seems amply supported. However, a demonstration of the truth of the first does not automatically justify the second. It is not at all clear what is the rhythmic process in the initiation of flowering.

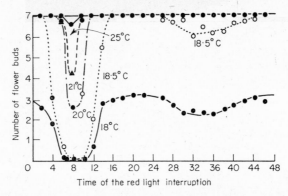

FIG. 5-7. The effect of temperature on the flowering of *Pharbitis nil* following a single dark period interrupted at different times with 5 minutes of red light, as shown on the abscissa. The periodicity is clearly independent of temperature although the number of flowers is increased by raising the temperature. (After Takimoto and Hamner, 1964.)

Although it probably concerns phytochrome action in some way, it does not seem to be a rhythmic change in the form of phytochrome itself. Cumming *et al.* illuminated *Chenopodium* seedlings at different times during a 72-hour dark period with mixtures of red and far-red light. Their results show that even a high proportion of far-red light, which would be expected to convert most of the phytochrome to the P_r form, did not promote flowering at times when red light alone was inhibitory. The rhythm in the effectiveness of red light could not be altered by changing the form of phytochrome P_{fr} giving promotion in the day phase and inhibition during the night phase whenever it was present. Such findings have led several investigators to postulate a rhythm in substrate concentration for the action of P_{fr}. However, there is as yet no definite evidence to confirm the suspicion that phytochrome must be an enzyme, and hence act on a substrate.

Is there any evidence that P_r as well as P_{fr} may be rhythmic in activity? Könitz demonstrated that, in *Kalanchoë*, exposure to red light at night was inhibitory to flowering as usual in short-day plants, but in addition, short exposures to far-red light were inhibitory during the day. There was an abrupt change from far-red to red inhibition at the end of 12 hours of light, where the transition from day to night would usually occur in nature. Cumming *et al.* demonstrated that, in *Chenopodium* seedlings, while red light, which would presumably change P_r to P_{fr}, showed cyclic inhibition of flowering, far-red light showed no such rhythm, but instead a gradually and steadily diminishing inhibition. Similar findings were also reported by Engelmann for *Kalanchoë*, by Salisbury for *Xanthium* and by Takimoto and Hamner for *Pharbitis*.

A fortuitously designed experiment of Hillman's with the tiny short-day plant *Lemna perpusilla* provides the best evidence that timing is inseparably connected to the phase of a circadian rhythm. *Lemna* can be grown in bacteria-free culture and fed with glucose so that it can be reared independent of requirements of light for photosynthesis. Thus long light periods can be eliminated, and 'skeleton' photoperiods, where short light pulses divide the long dark period into segments, can be used in experiments on the induction of flowering. In such experiments, Hillman found that *Lemna* will flower when given only two 15-minute light exposures in every 24 hours, the remainder of the time being in the dark and utilizing glucose as an energy source, if the light and dark treatments were arranged in the sequence: 15 minutes light, 13 hours darkness, 15 minutes light, $10\frac{1}{2}$ hours darkness. Further experiments showed that the sequence of these light and dark treatments is of the utmost importance. When the *Lemna* is transferred from the constant light under which the stock cultures are grown, the first dark period must be the *long* one for flowering to occur, that is the schedule must be: $13D:\frac{1}{4}L: 10\frac{1}{2}D:\frac{1}{4}L$, and not $10\frac{1}{2}D:\frac{1}{4}L:13D:\frac{1}{4}L$. It seems reasonable to conclude that the short light periods mark the beginning and end of two qualitatively different dark periods, one interpreted by the plant as day and the other as night. But can we tell which is which? If flowering occurs, the 13-hour dark period must have been interpreted as night. Therefore the first dark period and every alternate dark period thereafter is considered a night. This explains why no flowering occurred on the schedule which began with the shorter dark period. But why is the first dark period always interpreted as a night by *Lemna*? The behaviour of circadian rhythms explains this easily: like other plants in continuous bright light, *Lemna* is without a rhythm preceding the first dark period. This period initiates the oscillation of a circadian rhythm, and as always with the reinitiation of rhythmicity with a L-D transition out of bright continuous light, the rhythm starts with the beginning of the night phase. A 13-hour dark period will only act as a long night when it coincides with this night phase.

Hillman extended his experiments to confirm this interpretation by varying only the length of the first dark period after continuous light, and then giving a schedule of either $\frac{1}{4}$L:10$\frac{1}{2}$D:$\frac{1}{4}$L:13D or $\frac{1}{4}$L:13D: $\frac{1}{4}$L:10$\frac{1}{2}$D. As in the first experiments, promotion of flowering was altogether determined by the length of the *first* dark period. With the first schedule, flower induction occurred when this first dark period was 15 or 39 hours long, but not when it was 24 or 48 hours in duration. With the second light-dark schedule, most effective flower induction occurred with a preliminary dark period of 24 or 48 hours, just the reverse of the results with the first schedule.

The rhythmically changing effect of this first dark period followed by $\frac{1}{4}$L-13D:$\frac{1}{4}$L:10$\frac{1}{2}$D is shown in Fig. 5-8. These findings would be

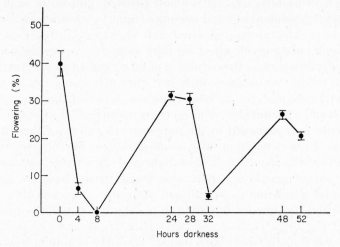

Fig. 5-8. The flowering of *Lemna perpusilla* treated with a dark period of different lengths preceding seven repetitions of the schedule $\frac{1}{4}$L:13D:$\frac{1}{4}$L:10$\frac{1}{2}$D. Twice the standard error of each mean is shown by brackets around each point. (After Hillman, 1964.)

very difficult to explain without invoking a circadian rhythm controlling time measurement, but fall into place nicely when circadian lore is taken into account, as follows: as in Hillman's first experiment, the *Lemna* stock cultures which have been kept in constant light for a long time are arhythmic. The rhythm is started by the first dark period, that is why it is the determining one. The phase of the rhythm is also determined by the beginning of this dark period as the beginning of a night phase. The 15-minute light exposures in the skeleton photoperiod are not sufficiently long or sufficiently close together to change the phase of the rhythm, which continues as initiated by the beginning of the first dark period. Flower promotion now occurs whenever a dark period longer than the critical night of *Lemna coincides with the night phase* of its circadian clock. Darkness during the day phase is ineffective in initiating flowering. Therefore it seems reasonable that this clock provides the time base to which the actual light and darkness of the environment is compared. The mechanism of comparison is still unknown and is another effect of light which does not involve a change of phase of the circadian rhythm, perhaps an effect mediated through phytochrome.

It is interesting to note that short photoperiods of the skeleton type, if properly arranged with respect to circadian time, will induce photoperiodic effects in insects (Minis, 1965, Pittendrigh, 1966, and Atkisson, 1966) and in birds (Menaker and Eskin, 1967). In these organisms, too, there is clear evidence that two kinds of light effects can change the photoperiodic response; one kind of light effect acts to determine the phase of the circadian clock and the other is the mechanism by which the coincidence of environmental light with the night phase of the circadian clock is sensed. Action spectra studies must be done to determine the nature of this second photoreaction.

Since the circadian clock is clearly implicated in photoperiodic time measurement, the design of future experiments on photoperiodism must take careful consideration of the phase relationships of circadian systems and the manner in which they can be reset. If investigators adhere to this stipulation, much confusion will be avoided and progress towards a more detailed understanding of the timing mechanism in photoperiodism should be rapid.

Evidence for the Existence of True Annual Rhythms

A rigorous demonstration of an annual rhythm in plants has probably never been achieved. The difficulties standing in the way of detecting such a rhythm are formidable. First, environmental conditions of light and temperature must be kept under strict control for at least three cycles, i.e. three years, not an easy task since equipment is subject to failure. Secondly, annual variation in the composition of media must be avoided. This poses almost insurmountable difficulty, since even twice-distilled water has been shown to vary in its organic composition, depending on whether or not the reservoir source of the water before purification contains algae or other organisms in bloom proportions. These blooms themselves tend to be seasonal. Photoperiodic control must be ruled out, when considering whether or not periodicities which appear to be annual in pattern are really annual rhythms in the strict sense. Even tropical plants, which grow in an environment which shows a minimum variation in day length with season, may be photoperiodically controlled. Recently, for example, the dropping of the leaves of the tropical shrub *Plumaria* have been shown to be under photoperiodic control (Murashige, 1966). Tropical forest trees do sometimes show cycles of defoliation where different trees in the same locality or even different branches of the same tree loose their leaves at different times. Responses to day length can hardly account for such behaviour. However, trees are intractible laboratory organisms and have only rarely been investigated under controlled conditions.

Small plants which can be grown under rigorously controlled temperature and unchanging light conditions have been observed to vary in growth rate over the year. Two investigators, Pirson and Göllner, 1953, and Bornkamm, 1966, have observed annual periodicity in *Lemna minor* L. under constant conditions. Pirson and Göllner measured the growth of *Lemna* roots, using as an indication the number of hours required for the roots to increase in length from 1 to 20 mm. In December, this required twice as long as in the previous or the following June. The decrease in the rate of root growth was closely correlated with increase in osmotic value of the root cells and a decrease in the time required to plasmolyse the cells. Changes in plasmolysis time were especially marked. In Bornkamm's experiments also the growth of *Lemna*, measured this time as increase in dry weight, was greater in summer although the plants were kept in constant light (1700 lux) and the temperature was maintained at 25° C. This occurred in two successive years, 1962 and 1963. Increased rate of dry weight accumulation was accompanied by an increase in the protein content and in the ratio of protein to carbohydrate. In the third year of Bornkamm's observations, however, no summer maximum was observed.

Annual variation in the growth rate of plant tissue cultures have also been noticed. For example, Shimon Lavee (personal communication) has observed that the cells of both dwarf and tall apple and of olive trees in culture show annual cycles of growth, measured either as wet or dry weight. Cultures were kept at 27–28° under constant light. The yearly cycles appeared in five successive years although the amplitude became somewhat reduced. Both this and the work with *Lemna* depend on growth on artificial media and hence are subject to error arising from undetected changes in the composition of the medium as mentioned above.

Neeb (1952) has reported that the alga *Hydrodictyon* is able to grow at higher light intensities during the late winter and spring than in the fall. The difference was quite distinct, 500 lux being tolerated in the spring, while this intensity killed cultures in the fall. The increase in sensitivity to light could be prevented if the plants were allowed a rest period of several weeks during which the algal nets were left closely packed together in not more than 250 lux. The meaning of these findings are hardly clear, especially since Neeb gives no quantitative data for the changes in sensitivity.

If there are annual differences in growth, it is not surprising that annual differences in the biochemical properties of cells can also be detected. Kessler and Czygan (1963) have made the interesting observation that intact cells of the green alga *Ankistrodesmus braunii*, in which the reduction of nitrite is inhibited by dinitrophenol, a seasonal difference in the capacity to reduce nitrate can be demonstrated, with maxima in June or July and minima in October and November in two successive years, although the cells were grown in constant light and temperature. The expected differences in activity could be demonstrated in extracts prepared from cells at different seasons. Using *Chlorella pyrenoidosa*, Kessler and Langner (1962) showed a similar annual variation in hydrogenase activity. Cultures were adapted to hydrogen for 20 hours and the reduction of nitrate with H_2 was then measured in the dark. Maxima were observed in late summer, the rate being five times that at the minimum in March. However, the cultures were grown in a window and hence were subject to annual fluctuations in light intensity and day length and no doubt also temperature.

The capacity of a number of kinds of seeds has been reported to vary over the year, when the seeds are kept air-dried and tested from time to time. Since it may be easier to keep the conditions of storage constant than to control the growth environment for non-dormant plants, seeds may be good material for detecting true annual rhythms. Bünning and Müssle (1951), and Bünning and Bauer (1952) report annual rhythms from studies of the germination of seeds of *Digitalis lutea*, *Chrysanthemum corymbosum* and *Gratiola officinalis*. The seeds of *Digitalis* and *Chrysanthemum* were stored at widely different temperatures, $-22°$, $4°$, $45°$ and $55°$ C, and samples were germinated at different times in the dark at $25°$ C. The storage temperature did not much affect the maxima and minima in the germination capacity, which occurred in early spring and late summer respectively. The gaseous atmosphere during storage (N_2, O_2 or CO_2) did not appear to affect the germination of *Gratiola* which germinated readily in January, February and March, but not at all during spring and summer. In no case were the seeds examined for a long enough time to observe more than two winter maxima in germination, however.

Ruge and Liedtke (1951) reported similar variation in the germination capacity of several species of *Malva*. The minimum occurred in April

D

with high germination from September to November. A sample of seeds from one species was tested through two full years and three winter maxima were found. However, the storage conditions of the seeds do not seem to have been controlled carefully. This may be of special importance with seeds of *Malva*, since the same authors have reported that heat treatments enhance germination and can induce good germination even at unfavourable times of year.

Bünning and Bauer reported that seeds of *Digitalis* with a low capacity to germinate seemed to have a somewhat lower water content and be more resistant to short exposure to high temperature than seeds with a higher germination. However, it is clear that the substantiation of the existence of annual rhythms in seeds and, in fact, in any plant material, will require further carefully controlled experimentation.

REFERENCES

Atkisson, P. L. (1966). Internal clocks and insect diapause. *Science N.Y.* **154**, 234–241.

Blaney, L. T., and Hamner, K. C. (1957). Interrelations among effects of temperature, photoperiod, and dark period on floral initiation of Biloxi soybean. *Bot. Gaz.* **119**, 10–24.

Bornkamm, R. (1966). Ein Jahresrhythmus des Wachstums bei *Lemna minor* L. *Planta* **69**, 178–186.

Borthwick, H. A., and Downs, R. J. (1964). Roles of active phytochrome in control of flowering of *Xanthium pennsylvanicum*. *Bot. Gaz.* **125**, 227–231.

Borthwick, H. A., Hendricks, S. B., Parker, M. W., Toole, E. H. and Toole, V. K. (1952). A reversible photoreaction controlling seed germination. *Proc. natn. Acad. Sci. U.S.A.* **38**, 662–666.

Bünning, E. (1936). Die endogene Tagesrhythmik als Grundlage der photoperiodischen Reaktion. *Ber. dt. bot. Ges.* **54**, 590–607.

Bünning, E. (1951). Über Langtagpflanzen mit doppelter photophiler Phase. *Ber. dt. bot. Ges.* **64**, 84–89.

Bünning, E. (1960). Circadian rhythms and the time measurement in photoperiodism. *Cold Spring Harb. Symp. quant. Biol.* **25**, 249–256.

Bünning, E., and Bauer, E. W. (1952). Über die Ursachen endogener Keimfähigkeitsschwankungen in Samen. *Z. Bot.* **40**, 67–76.

Bünning, E., and Moser, I. (1966). Response-Kurven bei der circadainen Rhythmik von *Phaseolus*. *Planta* **69**, 101–110.

Bünning, E., and Müssle, L. (1951). Der Verlauf der endogenen Jahresrhythmik in Samen unter dem Einfluss verschiedenartiger Aussenfaktoren. *Z. Naturf.* **6b**, 108–112.

Bünsow, R. C. (1953). Endogene Tagesrhythmik und Photoperiodismus bei *Kalanchoë blossfeldiana*. *Planta* **42**, 220–252.

Bünsow, R. C. (1960). The circadian rhythm of photoperiodic responsiveness in Kalanchoë. *Cold Spring Harb. Symp. quant. Biol.* **25**, 257–260.

Butler, W. L., Lane, H. C., and Siegelman, H. W. (1963). Nonphotochemical transformations of phytochrome *in vivo*. *Pl. Physiol. Lancaster* **38**, 514–519.

Carpenter, B. H., and Hamner, K. C. (1963). Effects of light quality on rhythmic flowering response of Biloxi soybean. *Pl. Physiol. Lancaster* **38**, 698–703.

Carr, D. J. (1952). A critical experiment on Bünning's theory of photoperiodism. *Z. Naturf.* **7b**, 570–571.

Claes, H., and Lang, A. (1947). Die Blütenbildung von *Hyoscyamus niger* in 48-stündigen Licht-Dunkel-Zyklen und in Zyklen mit aufgeteilten Lichtphasen. *Z. Naturf.* **2b**, 56–63.

Clauss, H., and Rau, W. (1956). Über die Blütenbildung von *Hyoscyamus niger* und *Arabidopsis* in 72-Stunden-Zyklen. *Z. Bot.* **44**, 437–454.

Coulter, M. W., and Hamner, K. C. (1964). Photoperiodic flowering response of Biloxi soybean in 72-hour cycles. *Pl. Physiol. Lancaster* **39**, 848–856.

Cumming, B. C. (1967). Circadian rhythmic flowering responses in *Chenopodium rubrum*: effects of glucose and sucrose. *Can. J. Bot.* **45**, 2173–2193.

Cumming, B. C., Hendricks, S. B. and Borthwick, H. A. (1965). Rhythmic flowering responses and phytochrome changes in a selection of *Chenopodium rubrum*. *Can. J. Bot.* **43**, 825–853.

Engelmann, W. (1960). Endogene Rhythmik und photoperiodische Blühinduktion bei *Kalanchoë*. *Planta* **55**, 496–511.

Furuya, M., and Hillman, W. S. (1964). Observations on spectrophotometrically assayable phytochrome *in vivo* in etiolated *Pisum* seedlings. *Planta* **63**, 31–42.

Garner, W. W., and Allard, H. A. (1920). Effect of relative length of day and night and other factors of the environment on the growth and reproducton in plants. *J. agric. Res.* **18**, 553–606.

Hamner, K. C. and Takimoto, A. (1964). Circadian rhythms and plant photoperiodism. *Am. Nat.* **98**, 295–322.

Hendricks, S. B. (1960). Rates of change of phytochrome as an essential factor determining photoperiodism in plants. *Cold Spring Harb. Symp. quant. Biol.* **25**, 245–248.

Hendricks, S. B. (1963). Metabolic control of timing. *Science N.Y.* **141**, 21–27.

Hillman, W. S. (1962). The Physiology of Flowering, pp. 42–53. Holt, Rinehart and Winston, New York, U.S.A.

Hillman, W. S. (1964). Endogenous circadian rhythms and the response of *Lemna perpusilla* to skeleton photoperiods. *Am. Nat.* **98**, 323–328.

Hillman, W. S. (1967). The physiology of phytochrome. *A. Rev. Pl. Physiol.* **18**, 301–324.

Hopkins, W. G., and Hillman W. S. (1965). Phytochrome changes in tissues of dark-grown seedlings representing various photoperiodic classes. *Am. J. Bot.* **52**, 427–432.

Hsu, J. C. S., and Hamner, K. C. (1967). Studies on the involvement of an endogenous rhythm in the photoperiodic response of *Hyoscyamus niger*. *Pl. Physiol. Lancaster*, **42**, 725–730.

Hussey, G. (1954). Experiments with two long-day plants designed to test Bünning's theory of photoperiodism. *Pl. Physiol. Lancaster* **7**, 253–260.

Kessler, E., and Czygan, C. F. (1963). Seasonal changes in the nitrate-reducing activity of a green alga. *Experimentia* **19**, 89–90.

Kessler, E., and Langner, W. (1962). Jahresperiodische Aktivitatsschwankungen bei einer Chlorella. *Naturwissenschaften* **49**, 331–382.

Könitz, W. (1958). Blühhemung bei Kurztagpflanzen durch Hellrot und Dunkelrotlicht in der photo- und scotophilen Phase. *Planta*, **51**, 1–29.

92   RHYTHMIC PHENOMENA IN PLANTS

Lörcher, L. (1958). Die Wirkung verschiedener Lichtqualitäten auf die endogene Tagesrhythmik von *Phaseolus*. *Z. Bot.* **46**, 209–241.

Menaker, M. and Eskin, A. (1967). Circadian clock in photoperiodic time measurement: a test of the Bünning hypothesis. *Science N.Y.* **157**, 1182–1185.

Minis, D. H. (1965). Parallel peculiarities in the entrainment of a circadian rhythm and photoperiodic induction in the pink boll worm (*Pectinophora gossypiella*). *In* Circadian Clocks (J. Aschoff, ed., pp. 333–343. N. Hollard Publishing Co. Amsterdam.

Murashige, T. (1966). The deciduous behavior of a tropical plant, *Plumaria acuminata*. *Pl. Physiol. Lancaster* **19**, 348–355.

Nanda, K. K., and Hamner, K. C. (1958). Studies on the nature of the endogenous rhythm affecting photoperiodic response of Biloxi soy bean. *Bot. Gaz*, **120**, 14–25.

Nanda, K. K., and Hamner, K. C. (1962). Investigations on the effect of 'light break' on the nature of the endogenous rhythm in the flowering response of Biloxi soybean (*Glycine max*, L. Mer.). *Planta* **58**, 164–174.

Neeb, O. (1952). *Hydrodictyon* als Objekt einer vergleichenden Untersuchung physiologischer Grössen. *Flora, Jena*, **139**, 39–95.

Papenfuss, H. D., and Salisbury, F. B. (1966). Light promotion of flowering and time measurement in *Xanthium*. *Z. Pfl. Physiol*. **54**, 195–202.

Pirson, A., and Göllner, E. (1953). Beobachtungen zur Entwicklungsphysiologie der *Lemna minor* L. *Flora, Jena* **140**, 485–498.

Pittendrigh, C. S. (1966). The circadian oscillation in *Drosophila pseudoobscura* pupae: a model of the photoperiodic clock. *Z. Pfl Physiol*. **54**, 275–307.

Ruge, A., and Liedtke, D. (1951). Zur periodischen Keimbereitschaft einiger Malven-Arten. *Ber. dt. bot. Ges*. **64**, 141–150.

Salisbury, F. B. (1959). Growth regulators and flowering. II: The cobaltous ion. *Pl. Physiol. Lancaster* **34**, 598–604.

Salisbury, F. B. (1963). Biological timing and hormone synthesis in flowering of *Xanthium*. *Planta* **59**, 518–534.

Salisbury, F. B. (1965). Time measurement and the light period in flowering. *Planta* **66**, 1–26.

Takimoto, A. and Hamner, K. C. (1964). Effect of temperature and preconditioning on photoperiodic response of *Pharbitis nil*. *Pl. Physiol. Lancaster* **39**, 1024–1030.

Takimoto, A. and Hamner, K. C. (1965a). Studies on red light interruptions in relation to timing mechanisms involved in the photoperiodic response of *Pharbitis nil*. *Pl. Physiol. Lancaster* **40**, 852–854.

Takimoto, A. and Hamner, K. C. (1965b). Effect of double light interruptions on the photoperiodic response of *Pharbitis nil*. *Pl. Physiol. Lancaster* **40**, 855–858.

Takimoto, A. and Hamner, K. C. (1965c). Effect of far-red light and its interaction with red light in the photoperiodic response of *Pharbitis nil*. *Pl. Physiol. Lancaster* **40**, 859–864.

Zimmer, R. (1962). Phasenverschiebung und andere Störlichtwirkungen auf die endogen tagesperiodischen Blütenblattebewegungen von *Kalanchoë blossfeldiana*. *Planta*, **58**, 283–300.

CHAPTER 6

Rhythms which do not match Environmental Periodicities

Beside the rhythms with periods that match the common oscillations of the environment, there is a whole array of phenomena which repeat at intervals of every imaginable length. In some the period is very short, as in beating flagella or cilia, and too fast for the eye to follow. In some it is so long that it is barely detectable in a human lifetime. In parts of Asia a man is said to be so old that he has seen the bamboo flower twice. The rhythms about which we are speaking here cannot be cued by the environment because it does not contain any variables with matching frequencies. Of course these periodicities are interesting in themselves, but they have an added attraction to students of circadian rhythmicity, since they may represent the raw material upon which evolution has acted to select those rhythms with periods which do approximate environmental periodicities.

It is outside the scope of this book and the ability of the author to describe every one of this multitude. Only a few examples will be selected to give an idea of how diverse they are. The very rapid beat of cilia and flagella may arise from alternate contraction of the fibre pairs on the two sides of the flagellum (Bradfield, 1955). It seems clear that ATP is involved in the contraction, since isolated flagella contract when this substance is added to the medium bathing them. What times the beat is not known. In addition, a wave of contraction may be seen to pass over the whole surface of organisms that are covered with cilia, so the timing of individual cilia must be coordinated and this coordination imposes a secondary rhythmicity on the beating. The rate at which the flagella beat increases as the temperature rises, at least in organisms where this has been investigated. Thus temperature-compensating mechanisms do not seem to have evolved, nor is any selective advantage of such a mechanism immediately obvious. In general, temperature dependence does not seem disadvantageous as long as rhythmicity is not serving the function of time keeping. The frequency of the ciliary rhythmic beat is of the order of 10–20 per second.

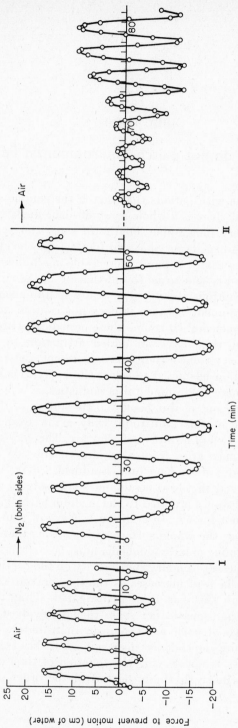

Fig. 6-1. The rhythmic changes in the direction of streaming in the plasmodium of *Physarum polycephalum*, measured as the amount of pressure necessary to prevent this motion, in air, then in nitrogen and finally returned again to air, at 20° C. (After Kamiya, Nakajima and Abe, 1957.)

Another rhythmic protoplasmic movement has been measured most ingeniously by Kamiya. This is the alternation in the direction of flow of the plasmodium of the slime mould, *Physarum polycephalum*. Plasmodia were persuaded to flow back and forth between two small chambers via a pore. The pressure within one of these chambers could be changed continuously so that the force was always just enough to prevent movement of the plasmodium. Beautiful records of the pressure applied in this way were obtained (FIG. 6-1). Furthermore, substances could be introduced into either chamber and their effect on streaming quantified. When both chambers were filled with N_2, the amplitude of the oscillation increased, while inspection of FIG. 6-1 shows that the period, about 3 minutes before anaerobiosis, was slowed to approximately 4 minutes by the addition of nitrogen. Adding 10^{-3}M KCN produced a change in both amplitude and frequency similar to that of N_2. Dinitrophenol appeared to produce quite a different effect from KCN, rhythmicity being very much slowed or abolished when this inhibitor was added to both chambers containing the plasmodium. Ether vapour also lengthened the period, while streaming in its presence appeared erratic. Addition of ATP (2×10^{-3}M) to one chamber increased the amplitude of the oscillation markedly but the period was slightly longer in its presence, 3 minutes as compared to 2 minutes before ATP was added, at 23° C. If ATP serves to provide the energy for streaming in *Physarum*, the results of the experiments with inhibitors and nitrogen certainly argue against the necessity of concurrent oxidative phosphorylation. Kamiya, Nakajima and Abe conclude that fermentation must be the source of ATP, a conclusion strengthened by the observation that iodoacetate, fluoride and dinitrophenol all interfere with streaming.

Kamiya (1953) measured the alternation of plasmodial streaming direction at a series of different temperatures to obtain information concerning the Q_{10} for both amplitude and frequency of this rhythm. Amplitude was only slightly affected by temperature, while the Q_{10} for the frequency was close to 2, thus the rhythmic aspect of plasmodial streaming shows marked temperature dependence.

The form of the oscillation is not a perfect sine wave, but its shape can be approximated by combining several sine waves. Different parts of a plasmodium may show oscillations differing in both phase and period. Nothing is known of the factors responsible for rhythmical alteration in the direction of streaming. However, unlike the oscillations represented by ciliary beat, the near sinusoidal nature of the rhythm in direction of streaming argues against a relaxation oscillation as the cause.

An electrical potential has been found to exist between the tip and the base of roots in dilute salt solutions. In the bean root studied by Scott (1957), small apparently spontaneous oscillations with periods of about 5 minutes can sometimes be found in this potential. These may persist for hours. Similar rhythms can be induced by stimulating the root electrically, mechanically or even chemically, but these damp out very quickly as a general rule. When the root was alternately placed in water, and 0·033M sucrose, the electrical potential changed in response to the different osmotic potential of the medium, but the changes were much greater when the medium was changed at about 5 minute intervals than at longer or shorter times, suggesting that 5 minutes represented the natural resonant frequency of potential. Roots responded to repetitive changes in auxin level in the medium in a similar manner. After such treatment, the continuing oscillations were often more pronounced than in the absence of external oscillations of the resonant frequency. These observations led Scott to postulate hunting in a loop of a feedback control mechanism in the root as the cause for the rhythms seen both spontaneously and after stimulation.

Blinks has observed somewhat similar oscillations in the electrical potential measured between electrodes inside and outside the large coenocyte of *Halicystis*, the gametophytic stage of the green alga, *Derbesia*, the same organism in which the question of semilunar rhythms in gametogenesis has been examined (see Chapter 4). These oscillations may be induced, for example, by the addition of potassium, and they oscillate with a frequency of 10 minutes, but damp out rapidly. Lighting or electrical stimulation can cause the rhythmic behaviour to recommence.

The induction of short period rhythms not present in the absence of some sort of stimulation may be detected chemically as well as electrically. Betz and Chance (1965) have shown that when yeast cells (*Saccharomyces carlsbergensis*) are grown aerobically and then transferred to anaerobic conditions with the addition of sugar, the amount of reduced pyridine nucleotide oscillates. Observations of the phenomenon were made with a double beam spectrophotometer capable of determining the changes in pyridine nucleotide absorption with reduction and oxidation so that very rapid changes could be followed. The oscillations which they could observe in this way were sinusoidal, although they sometimes varied considerably from simple sine waves. The period *in vivo* at 25° C was about 0·6 minute, and was markedly temperature dependent, one preparation showing a period of 37 seconds at 25° C and 15 seconds at 35° C. The oscillations were strongly damped in many cell preparations. In parallel with the reduction of pyridine nucleo-

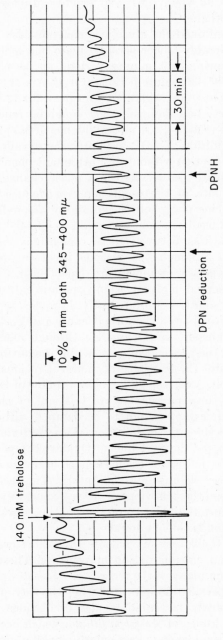

FIG. 6-2. Oscillations in the reduction level of pyridine nucleotide in yeast extracts, sustained by the addition (at arrow) of 140 mM trehalose. The period is approximately 7 minutes at 25° C. Reproduced with permission from Pye and Chance (1966). *Proc. natn. Acad. Sci. U.S.A.*

tide were decreases in ATP, glucose-6-phosphate and fructose-6-phosphate, and increases in ADP, AMP and fructose diphosphate, all of which oscillated together.

Extracts which contained the glycolytic enzymes also oscillated *in vitro* but with considerably longer periods, again markedly dependent on temperature, showing 7–9 minute periods at 25° C, 41 minute periods at 12° C and 86–98 minute periods at 0° C. About 8 cycles could be seen clearly. Thus at 0° C oscillations continued for 12 hours. When oscillations in extracts had damped, they could be renewed by the addition of trehalose (FIG. 6-2). A phase advance of 3·5 minutes could be induced by the addition of ADP just after the pyridine nucleotide reached maximum reduction without any attendant change in period. The addition of pyruvate also resulted in a phase change, this time a delay. The primary site of the oscillation was deduced to be the activation of the enzyme phosphofructokinase by its product, fructose diphosphate, hence a feedback loop was again implicated as the cause of oscillation. Similarities between these oscillations and circadian rhythms were clear, although the periods involved were of very different magnitude and were temperature dependent.

The oscillations observed in glycolysis in yeast and extracts prepared from it take place in solution. More recently, however, similar oscillations have been produced in mitochondria (Chance and Yoshioka, 1966). These involve the transport of H^+ and K^+ through membranes and they are generated in the presence of oxygen and the antibiotic valinomycin, which stimulates the transport of monovalent ions such as K^+ into the mitochondrion. A corresponding outward flow of H^+ also takes place. Oscillations in these parameters are 180° out of phase or very nearly so, and sometimes appear not to be damped, although this is difficult to determine since the exhaustion of oxygen via respiration soon abruptly terminates the periodicity. Oxidative processes in the mitochondrion must be intact for oscillations to occur, and ATP may be a prerequisite since oligomycin is inhibitory. The lack of precise correspondence between the behaviour of K^+ transport as such and the oscillation argue against the identity of the two processes. The exact nature of the oscillation is thus in doubt.

Leaves often show oscillatory movements with short periods, sometimes superimposed on circadian sleep movements. These appear as series of small spikes on many of the records of *Phaseolus* published by Bünning, and are especially prominent during the day phase of the circadian rhythm. Darwin measured regular oscillations with a 4–5 minute period in the leaves of *Averrhoa bilimbi* which occurred while they were closing in the evening. Particularly striking are the move-

ments of the leaves of *Desmodium gyrans* from which the specific name
of this plant is derived. The up and down motion of the leaflets of *Desmodium* have been recorded and appear very regular for at least four
hours, oscillating with a period of about 3 minutes (Bose, 1927).
Guhathakurta and Dutt showed that electrical oscillations accompanied
the changes in leaf position in *Desmodium*. Movements of leaves are
brought about by osmotic changes in the cells of the pulvini and subsequent alterations in turgor. Interesting for comparison is the occurrence
of rhythmicity in the degree of opening of stomata in *Vicia faba* reported
by Stälfelt. The period of this rhythm was 15–20 minutes. While there is
considerable variation in single determinations of the period of this
rhythm, the average length seems to be similar at 20° C and 30° C. The
change in the opening of the stomata was accomplished by the swelling
and shrinking of the guard cells which appeared to be synchronized.
Recently it has been shown that the changes in turgor of guard cells
require the presence of potassium, and can be correlated with electrical
potential changes, implicating the potassium pump, possibly thus the
site of rhythmicity (Fischer, 1968).

Rapidly growing portions of plants, like the shoot and root tips of
seedlings and young tendrils, do not usually grow in a constant direction.
When careful records are made of their progress, they are seen to wave
back and forth describing with their tips circles or eclipses of various
shapes. Baillaud has observed the stems of *Cuscuta* and reports that
they repeat this pattern with a period of 5–10 minutes at 23° C. The
movements are very temperature dependent, the period showing a Q_{10}
of about 2. Galston *et al.* utilized a time-lapse camera to observe pea
seedling plumules. When completely etiolated, plants showed little
movement, but in continuous red light, the plumules rotated with a
period of 77 minutes (FIG. 6-3).

Spurný (1968) reports spiral oscillations in the growing radicle of
pea. Hypocotyls of safflower (*Carthamus tinctorius*) also oscillate but
only as a result of geotropic or phototropic stimulation. The period is
not very temperature dependent, being 240 minutes at 20° C and 210 at
30° C (Karvé and Salanki, 1964). A second geotropic stimulus initiates a
new phase, especially when the stimulus is given at or a little after the
maximum of the previously-induced rhythm.

Fungi show rhythms in growth which under certain circumstances
appear to have periods of 24 hours, but which have properties which set
them apart from typical circadian rhythms. Two growth forms alternate
when the fungus is inoculated at one end of a long tube partly filled with
solid medium and grown in the dark. Unlike circadian rhythms, the
period depends very markedly on the temperature. Periods as long as

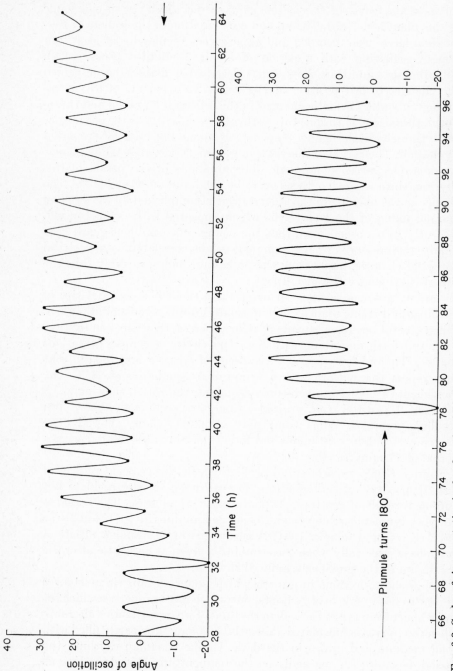

FIG. 6.3. Cycles of circumnutation in the plumule of *Pisum sativum* L. cv. 'Alaska' germinated in darkness and transferred to constant red light of low intensity (330 ergs/cm²/sec) before hour 28. Period 77 minutes, at room temperature (26° C). (From Galston, Tuttle and Penny, 1964).

90 hours can be observed. Furthermore, the rhythm in a 'clock' mutant of *Neurospora crassa* is not reset by an exposure to light (Berliner and Neurath, 1965). The phase is established by the time of inoculation into new medium. Grown in the dark on a medium containing yeast extract and a number of amino acids, *Aspergillus ochraceus* and *Colletotrichum Lindemuthianum* display a 4 day periodicity in growth pattern at 23° C (Jerebzoff, 1965). Minor constituents in the medium, particularly amino acids, seem to make the difference between zonation and none. These rhythms are not to be confused with the truly circadian growth rhythm observed in *Neurospora crassa* 'timex' mutant (Sargent and Briggs, 1967). The existence in fungi of rhythms, similar in many respects to circadian rhythms, yet capable of a much longer period than 24 hours, temperature dependent and not resetible by light, is most interesting because they might be considered to represent the ancestral type from which the more sophisticated circadian organization has evolved. This view is strengthened by the presence of truly circadian rhythms in many fungi, even a mutant of the same species of *Neurospora* which shows the less differentiated type of rhythm.

Rhythms with periods of months or years are understandably poorly studied. That they do exist is indicated by at least a few observations made under controlled conditions. For example, Lavarenne-Allary (1966) has grown seedlings of several species of oak in controlled environments in LL at 27° C. These seedlings alternately grow and become dormant, as do tree seedlings in nature, showing 4–8 such periods in the course of 7 months. In onion bulbs which were stored at 5° C for two years and sampled twice each month during this time, Jaffe and Isenberg (1965) report a periodicity in rate of elongation of excised leaves, the period of which was 2·5 months.

While experiments under controlled conditions have not been attempted, observations in the field indicate that the large tropical bamboos in India and Jamaica flower at intervals of thirty or forty years. That this is not the result of environmental cues is indicated by the simultaneous flowering of bamboo derived from the same stock in very different localities, the gardens at Kew and the tropical rain forests of Jamaica for example (Young and Haun, 1961). Walker (1956) reported that the *Laminaria* population off the coast of Scotland varied in density over a ten year period but this variation showed a strong correlation with sun spot activity and thus weather changes are probably responsible, and the variations are not rhythms. The understanding of factors controlling very long cycles, whether exogenous or endogenous, must await long-term cooperative investigations.

REFERENCES

Baillaud, L. (1953). Action de la température sur la période de nutation des tiges volubiles de Cuscute, *C. r. hebd. Séanc. Acad. Sci., Paris* **236**, 1986–1988.

Berliner, M. D., and Neurath, P. W. (1965). The band forming rhythm of *Neurospora* mutants. *J. cell. comp. Physiol.* **65**, 183–194.

Betz, A., and Chance, B. (1965). Phase relationship of glycolytic intermediates in yeast cells with oscillatory metabolic control. *Archs Biochem. Biophys.* **109**, 585–594.

Bose, J. C. (1927). Plant autographs and their revelations. 231 pp. Longmans, Green and Co., London.

Bradfield, J. R. G. (1955). Fibre patterns in animal flagella and cilia. *Symp. Soc. exp. Biol.* **9**, 306–331.

Chance, B., Eastbrook, R. W., and Ghosh, A. (1964). Damped sinusoidal oscillations of cytoplasmic reduced pyridine nucleotide in yeast cells. *Proc. natn Acad. Sci. U.S.A.* **51**, 1244–1251.

Chance, B., Schoener, B., and Elsaesser, S. (1964). Control of the waveform of oscillations of the reduced pyridine nucleotide level in a cell-free extract. *Proc. natn Acad. Sci. U.S.A.* **52**, 337–341.

Chance, B., and Yoshioka, T. (1966). Sustained oscillations of ionic constituents of mitochondria. *Archs Biochem. Biophys.* **117**, 451–465.

Fischer, R. A., (1968). Stomatal opening: role of potassium uptake of guard cells. *Science, N.Y.*, **160**, 784–785.

Galston, A. W., Tuttle, A. A., and Penny, P. J. (1964). A kinetic study of growth movements and photomorphogenesis in etiolated pea seedlings. *Am. J. Bot.* **51**, 853–858.

Guhathakurta, A., and Dutt, B. K. (1961). Electrical correlate of the pulsatory movement of *Desmodium gyrans*. *Transl. Bose Res. Inst.* **24**, 73–82.

Jaffe, M. J., and Isenberg, F. M. R. (1965). Rhythmic growth in excised sprout-leaves of onion bulbs. *Pl. Physiol. Lancaster*, **40**, *supplement* xx (abstract).

Jenkinson, I. S., and Scott, B. I. H. (1961). Bioelectric oscillations of bean roots: further evidence for a feedback oscillator. I. *Aust. J. biol. Sci.* **14**, 231–247.

Jenkinson, I. S. (1962). Bioelectric oscillations of bean roots: further evidence for a feedback oscillator. II. *Aust. J. biol. Sci.* **15**, 101–114.

Jenkinson, I. S. (1962). Bioelectric oscillations of bean roots: further evidence for a feedback oscillator. III. *Aust. J. biol. Sci.* **15**, 115–125.

Jerebzoff, S. (1965). Manipulation of some oscillating systems in fungi by chemicals. pp. 183–189 *In* Circadian Clocks (J. Aschoff, ed.) North Holland Publishing Co. Amsterdam.

Kamiya, N. (1953). The motive force responsible for protoplasmic streaming in the myxomycete plasmodium. *Ann. Report. Faculty of Science, Osaka University*, **1**, 53–83.

Kamiya, N., Nakajima, H., and Abe, S. (1957). Physiology of the motive force of protoplasmic streaming. *Protoplasma* **48**, 94–112.

Karvé, A. D., and Salanki, A. S. (1964). The phenomenon of phase shift in geotropically induced feedback oscillations in *Carthamus tinctorius* L. seedlings. *Z. f. Bot.* **52**, 113–117.

Lavarenne-Allary, S. (1966). Croissance rythmique de quelques espèces de Chêne cultivées en chambres climatisées. *C. r. hebd. Séanc. Acad. Sci., Paris* **262** D; 358–361.

Pye, K., and Chance, B. (1966). Sustained sinusoidal oscillations of reduced pyridine nucleotide in cell-free extract of *Saccharomyces calsbergensis*. *Proc. natn Acad. Sci. U.S.A.* **55**, 888–894.

Sargent, M. L., and Briggs, W. R. (1967). The effect of light on a circadian rhythm of conidiation in *Neurospora*. *Pl. Physiol. Lancaster*, **42**, 1504–1510.

Scott, B. I. H. (1957). Electric oscillations generated by plant roots and a possible feedback mechanism responsible for them. *Aust. J. Biol. Sci.* **10**, 164–179.

Scott, B. I. H. (1962). Feedback-induced oscillations of five-minute period in the electric field of the bean root. *Ann. N.Y. Acad. Sci.* **98**, 890–900.

Spurný, M. (1968). Spiral oscillations of the growing radicle in *Pisum sativum* L. *Naturwissenschaften*, **55**, 46.

Stålfelt, M. G. (1929). Pulsierende Blattgewebe. *Planta* **7**, 720–734.

Stålfelt, M. G. (1965). The relation between the endogenous and induced elements of the stomatal movements. *Physiol. Pl.* **18**, 177–184.

Walker, F. T. (1956). Periodicity of the Laminariaceae around Scotland. *Nature Lond.*, **177**, 1246.

Young, R. A., and Haun, J. R. (1961). Bamboo in the United States: description, culture, utilization. Agricultural Handbook No. 193.

CHAPTER 7

The Cell Division Cycle

All cells show another periodic phenomenon which we have not yet considered—the cyclic change between vegetative existence and cell division. When large numbers of cells are associated in a tissue or a population of single cells, mitosis of individuals is often not synchronized. It is not possible to detect differences associated with the cell division cycle in such populations because at any given time, there are members of the population in all stages of the cycle. However, the study of single isolated cells has shown that not only is there a sharp differentiation between dividing and non-dividing cells, but the vegetative life of a cell can be further subdivided into stages which follow each other in an orderly sequence. A newly formed daughter cell enters a rest period as far as reproduction is concerned, which has been named the first gap or G_1 phase. During this time no new DNA is synthesized and the nucleus appears to be engaged in activities not related to cell division. This is followed by the S period, during which new DNA is synthesized to double the DNA content of the cell. Usually mitosis does not follow immediately after the replication of DNA, but another gap in processes associated with cell division can be distinguished, which is called G_2. Mitosis proper, with the formation of the mitotic spindle, the separation of homologous chromosomes and cytokinesis, then follows to complete the cell cycle. Different kinds of cells vary in the time they spend in any one phase of the cell cycle, and may lack distinct gaps, as do bacteria.

The cell cycle can be studied either in isolated cells or in cultures in which cell division has been synchronized in some way. The latter are necessary for most kinds of biochemical work because large amounts of material are required. Consequently, considerable research has been directed toward the development of methods by which the cells of a population can be synchronized with respect to division. Manipulations which temporarily prevent cells from initiating division, e.g. brief exposure to high or low temperature or temporary withdrawal of a necessary nutrient, have sometimes been successful. Lewin *et al.* (1966) have shown that removal of silicon from cultures of the diatom *Navicula pelliculosa* arrests all the cells at a stage prior to cytokinesis. On the re-admission of silicon, all then divide in synchrony. In photosynthetic organisms, a similar result may be achieved by keeping the culture in darkness for a long time, then illuminating. Von Denffer (1949) was able to synchronize cultures of the diatom *Nitzschia palea* in this way, by darkening them for 10 days, then illuminating them continuously. However, synchrony was not complete and died away after two bursts of division. Lewin *et al.* were also able to synchronize cultures of a diatom, *Cylindrotheca fusiformis*, with a long dark period followed by light.

Some photosynthetic cells can be synchronized by growing them in a light-dark cycle. In general, when cells are synchronized in this way, division occurs during the dark part of the cycle.

If synchrony is to be complete in LD, the light part of the cycle must be long enough for the completion of light-requiring processes which are necessary for cell division. Cook and James (1960) offer evidence that 13·7 hours is long enough for this purpose in *Euglena*, since cells exposed to light for this length of time will subsequently divide during the next 6·2 hours in darkness. Under different light conditions, Edmunds finds 12 hours of light sufficient for cell division in *Euglena* (FIG. 7-1). Lee-dale's experiment (1959) with a number of species of *Euglena* have suggested that inhibition of cell division during the light period may also contribute to the development of synchrony. When grown in unen-riched soil-water culture in natural light, *Euglena* divides only during the night, the maximum number of cells undergoing division being found at midnight. No divisions at all occurred in continuous light or continuous darkness. If cultures were illuminated after they had been in darkness less than an hour, they did not divide. If some hours of darkness intervened, however, then cell division could be completed in light. In organic media, *Euglena* is quite able to divide in continous light, with no evidence that this light causes inhibition. A number of investigators have found that the addition of even small amounts of organic substrates like beef extract, ethanol (FIG. 7-1) or acetate to cultures of *Euglena* lead to immediate loss of synchrony in previously synchronized cultures, even in a light-dark cycle. In this connection, it is interesting to note that colourless euglenids cannot be synchronized with a light-dark cycle, either natural or artificial, although *Astasia* can be synchronized by a cycle of high and low temperature.

Cell division in a number of autotrophic cells beside *Euglena* becomes synchronous in light-dark cycles. Synchrony has been successfully achieved in this way in several species of diatoms, both pennate and centric, perhaps first by Rieth (1939) in *Biddulphia* and *Coscinodiscus*, and by von Denffer (1949) using *Nitzschia palea*. In the latter diatom

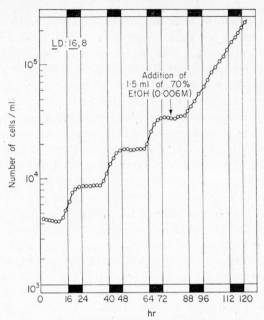

FIG. 7-1. Synchrony of cell division in *Euglena gracilis* (Z strain) induced by light-dark cycles (LD, 16:8, 25° C), and its abolishment by the addition of ethanol. (From Edmunds, 1965. *J. Cell comp. Physiol.* **66**, 147–158).

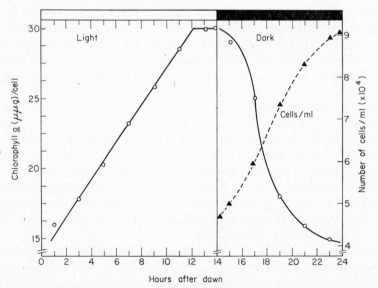

FIG. 7-2. Change in the content of chlorophyll *a* during the cell cycle in synchronized cultures of *Euglena gracilis* (Z strain), L:D, 14:10, 3500 lux, 25° C. (From Edmunds, 1965b. *J. Cell. comp. Physiol.* **66**, 159–182.)

the shortest generation time attained was about 16 hours and light-dark cycles LD 8 : 8̄ led to good synchrony in division. If the cycle was shortened to LD 6 : 6̄, division occurred every second dark period. *Skeletonema*, too, can be synchronized with a LD 12 : 1̄2̄ (Jørgensen, 1966) although division is spread over some 9 hours and starts in the middle of the light period.

Processes other than those directly connected with cell division may be restricted to certain stages in the cell cycle. In *Euglena* (Cook, 1961, Edmunds, 1965b) the synthesis of chlorophyll *a* and the carotenoids is completed during the first 12 hours of the synchronizing 16 : 8̄ LD cycle,

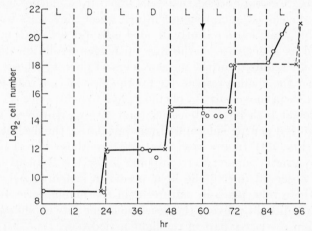

FIG. 7-3. Synchrony in cultures of *Chlamydomonas moewusii* in LD, 12:12, 8000 lux 25° C, and then in LL after hour 60. (After Bernstein, 1964.)

(FIG 7-2) while the synthesis of the cell protein as a whole occurs somewhat earlier. In *Skeletonema* also, increase in chlorophyll *a* and *c*, and fucoxanthin occurs only during the light part of the cycle. Photosynthesis begins to increase during the dark period, however, about 4 hours after the appearance of new cells.

Synchrony in reproduction can also be achieved in algal cells which, instead of dividing by simple fission, give rise to a number of daughter cells at one burst. Two such organisms have been studied in some detail, *Chlorella* and *Chlamydomonas*. The burst in *Chlamydomonas* can be especially well synchronized by light-dark cycles (LD 12 : 1̄2̄), so that all daughter cells appear at once in one hour at the end of the dark period (Bernstein, 1960, 1964, FIG. 7-3). With both *Chlamydomonas* and *Chlorella*, light-dark cycles are supplemented by renewing the culture medium at the beginning of the light period, thus keeping the culture in the log phase of growth.

Striking segregation of physiological processes has been detected in synchronized cultures of *Chlorella* by Tamiya and his co-workers (Tamiya, 1967), by Sorokin (1960, 1963) and by Pirson and Lorenzen (1958). Photosynthesis in particular shows variations in rate which can be correlated with the cell cycle. Sorokin, working with the high temperature strain of *Chlorella pyrenoidosa* 7-11-05, synchronized at 39° C and 10,000 lux with a LD 9:15̄, found that less oxygen was evolved by a cell in the light at the beginning and the end of the light period than during the second hour of illumination. In the course of measurements of oxygen evolution, which were made using the manometric technique of Warburg, Sorokin noticed that cells taken from culture later in the cell cycle developed symptoms of injury when they were exposed to bright light, while more recently divided cells did not. This injury was manifested as a decrease in the rate of oxygen evolution during the several hours required for the measurement of photosynthesis. He considered that the change responsible for this destructive reaction might be an increase in the sensitivity of chlorophyll to light, since the decrease was greater at high light intensities than at low. However, when the oxygen evolved per mg chlorophyll was calculated, the decline was still observed (FIG. 7-4). In any case, it is clear from these data that changes in photosynthetic capacity are responsible for the changes in photosynthesis observed during the light period in synchronized *Chlorella*, whether these take place in culture or in the manometric apparatus. Bicarbonate did not prevent the decrease in the rate of photosynthesis in cells from the latter part of the light period.

In considering the effects of various factors such as light and temperature on the cell cycle in such organisms as *Chlorella*, it must be remembered that the cell does not divide by simple fission. Instead, a number of new cells emerge at each division burst, and this number is by no means constant. The response to more favourable conditions may be an increase in the number of daughter cells produced at one burst, rather than the more frequent production of these bursts. When the data are given as doubling times, no distinction is made between these possibilities. In considering the studies of Sorokin on the effect of temperature and light intensity on the accumulation of cell substance during a light period in synchronized *Chlorella* cultures, it must be remembered that the 'doubling time' does not have anything to do with the number of cells dividing, but reflects just the total amount of cell material which is synthesized. This synthesis takes place exclusively during the time that the cells are in the light.

Pirson and Lorenzen (1958) were able to obtain good synchrony of the Emerson strain of *Chlorella pyrenoidosa* with light-dark cycles

(LD 16:12, 9000 lux, at 30° C) if new medium was added at the beginning of each light period. The rate of photosynthesis in these cells rose
to a maximum in the first part of the light period, about hour 8 (FIG.
7-5). At this time, too, the cells are most sensitive to damage by short
exposure to low temperature.

Kates and Jones (1967) have studied biochemical differentiation in
the cell cycle in synchronized cultures of *Chlamydomonas reinhardti* and

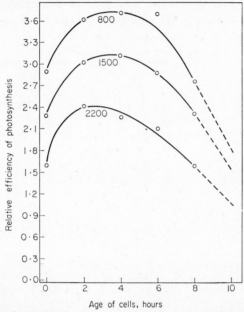

FIG. 7-4. The relative efficiency of photosynthesis in synchronized *Chlorella* strain
7-11-05 at different developmental stages at different light intensities, as given in lux on
each curve. The relative efficiency is expressed as mm³ O_2 per relative unit chlorophyll
per lux per hr. (After Sorokin and Krauss, 1961.)

C. moewusii. In particular, they examined the activity of the enzymes
which are implicated in the incorporation of ammonium ions into amino
acids. The activity of two enzymes varied with time, reaching a maximum about 6–9 hours after suspension of the cell in new medium. It is
interesting to note that the activity of malic dehydrogenase did not
vary.

Although most cell populations are not naturally synchronized even
when they are growing in a light-dark cycle, there are certain tissues
where natural synchrony of cell division can be found. In the develop-
ing anthers of *Lilium* and *Trillium*, for example, meiosis and the mitosis
which follows occur synchronously. The development of the anthers
takes place slowly and the stage at any given time can be determined

with fair accuracy by measuring the length of the bud. This situation has been exploited by Stern and his collaborators to investigate the biochemical changes which precede nuclear events both in meiosis and mitosis. They found that the enzyme thymidine kinase is present in Lily microspores only for a relatively short time during the interval between meiosis and the first postmeiotic mitosis, that is for only 24 hours in this 2-week interval. Its appearance can be prevented by pretreatment with inhibitors of protein synthesis. Inhibitors of RNA administered somewhat earlier also inhibit what appears to be the *de novo* synthesis of this

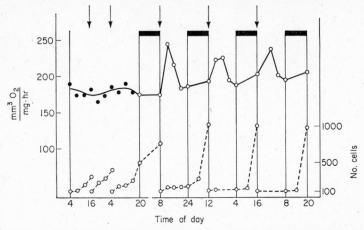

FIG. 7-5. Synchrony in reproduction (lower curve) and in photosynthesis (upper curve), in *Chlorella pyrenoidosa*, Emerson strain, in LD, 16:12, 9000 lux, 30° C, measured as O_2 mm^3 evolved in 20 min. at 15,000 lux and 30° C per mg., dry weight. (After Pirson and Lorenzen, 1958.)

enzyme. The rapid disappearance of thymidine kinase also requires a control mechanism. Two distinct maxima in protein synthesis accompany meiosis. Stern (1961) found that, during the development of the anthers, DNase appeared and disappeared a number of times. However, the intervals were irregular rather than rhythmic.

It is clear then from studies of both artificially and naturally synchronized cells that different parts of the cell cycle are characterized by different biochemical and physiological processes and specific enzyme activity may be limited to only a small part of the cell cycle. The enzymes concerned may be part of the machinery for the synthesis of DNA in the S phase, but they may also function in quite different processes, for example in pigment synthesis or photosynthesis. The contribution which the division of chloroplasts and mitochondria make to biochemical or functional changes during the cell cycle is not yet known but must be important.

Perhaps you have noticed that the light-dark cycles most commonly used to synchronize cell division in cultures of various algae have a total length of 24 hours. This is hardly a coincidence. Perhaps this observation is only a reflection of the fact that investigators are operating on a daily schedule and so such LD schedules fit their work habits. Then, too, the most easily available time clocks for setting up LD schedules in the laboratory run in 24-hour cycles so that it is difficult if not impossible to arrange for LD cycles which do not add up to a total of 24 hours. However, there may be a deeper significance to the prevalence of 24-hour cycles of LD as synchronizing regimes. All the organisms with which we work have evolved in a strong environmental cycle with a 24-hour period, and selection may well have operated to produce cells which time their division cycle to fit this external periodicity. The crucial question is whether or not the cell utilizes the same endogenous clock which controls circadian rhythms in other functions to time the division cycle as well. Bruce (1965) has discussed this possibility in 'Circadian Clocks'. He points to the common occurrence of generation times about 24 hours long but also calls attention to the observation that in many cells the time at which division takes place does not bear any fixed relationship to the light-dark cycle. This is especially true in animal cells where cell division may occur every 24 hours but is timed by other factors like the time of wounding. In plant cells also a 24-hour period may be observed but it has been shown that the time of division is related to the time of application of a hormone, for example gibberellin (Sachs et al., 1959). It has even been suggested that the division cycle is itself the basic circadian oscillator and all other functions which show this period are controlled by it, the LD cycle acting as a synchronizing agent. How can we distinguish between these alternatives?

First, does cell division have the properties of a circadian rhythm? If it is the basic oscillator responsible for controlling all other rhythms with this period, then cell division itself should show the temperature independence typical of circadian ryhthms. In many organisms, the time required for the completion of one cell cycle—the generation time of a cell—increases greatly as the temperature is lowered below the optimum for the cell in question, and thus shows a marked temperature dependence. Lowering the light intensity also results in a longer cell generation time, while the period of circadian rhythms is only slightly affected. Dinoflagellates quite commonly divide in partial synchrony when they are cultured in an LD cycle (FIG. 7-6). However, on careful examination, this synchrony appears different from that in *Chlorella*, the diatoms and some other organisms. First, cell division continues to be rhythmic in LL for many days although only a small part of the population is dividing at any one burst. Then, too, in *Gonyaulax*, lowering the temperature below 20–25° C. increases the generation time but the time between the successive division burst is actually *decreased* just as is the time between maxima in luminescence, and to the same extent. The number of cells which divide at each burst, however, is decreased and so the *average* generation time of a culture is increased by lowering the temperature. The time when these cells divide thus appears to be controlled by a mechanism outside the division cycle itself, in such a way that there is a certain 'allowed time' for cell division to take place

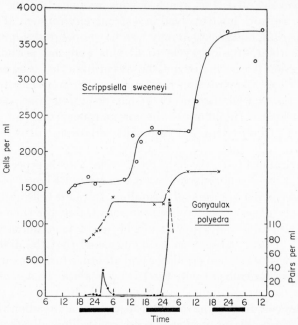

Fig. 7-6. Growth of cultures of *Scrippsiella sweeneyi* in natural light (upper curve) and *Gonyaulax polyedra* in LD, 12 : 12 (middle curve) and the number of paired cells in the same culture of *Gonyaulax* (lower curve). After dividing, cells remain attached to form a pair for about 30 minutes. (From Sweeney and Hastings, 1964.)

in each circadian cycle. If a cell is not ready to divide at this time, it must wait for the next allowed time before dividing. As a result, not all generation times are possible. That this is the case is clear when the generation time is measured for cells separately rather than as a population. Such experiments were carried out with several dinoflagellates. When cells are isolated in a capillary and the time from one division to the next is measured, a frequency diagram can be plotted (FIG. 7-7) which clearly shows that the generation time may change but does so discontinuously so that it is always about $n \times 24$ hours.

Furthermore, circadian rhythms in luminescence and photosynthesis in *Gonyaulax* are present in populations which are not dividing, either because they have reached the plateau of the growth curve or because the light intensity is too low, as it is in many experiments in constant dim light where the intensity is just above the level of compensation of photosynthesis.

In *Acetabularia*, too, a circadian rhythm can easily be demonstrated although the nucleus is not dividing. Even cells without a nucleus are

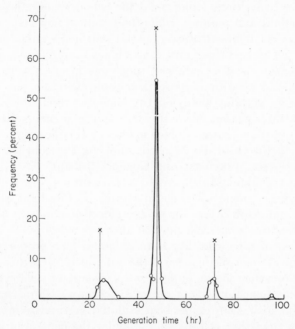

FIG. 7-7. The frequency of different generation times in single isolated cells of the dino-flagellates *Gonyaulax polyedra* (x) and *Gymnodinium splendens* (○) in LD, 12:12.

rhythmic, so that a cell cycle can clearly not be responsible for timing the circadian period.

Considering these findings, it is hardly feasible to invoke the cell cycle as the basic oscillator responsible for circadian rhythms in general or even cell division rhythms like that in *Gonyaulax*.

Perhaps the reverse is generally true: cell division is usually under the control of the circadian oscillator and this may be why LD cycles of 24 hours duration are successful in bringing about synchrony. For synchrony to occur in a population of *Gonyaulax* it would only be necessary to find environmental conditions where all cells can prepare for division in a single day. Unfortunately, such conditions have not been achieved with this cell, the shortest generation time usually turning out to be 48 hours on any LD cycle in which the dark period is long enough to be perceived as different from constant light. In very short dark periods and in LL, generation times may be shorter but become randomly distributed in time, in keeping with the generalization that the circadian rhythms in *Gonyaulax* disappear in bright LL. There is some evidence that in other cells which can be synchronized, the circadian oscillator may be in control of division. Lorenzen has found that when *Chlorella* is grown under conditions suboptimal for cell division, the number of daughter cells is first reduced, while the bursts of new cells continue to appear every 24 hours. However, under still more unfavourable conditions the generation time jumps to 48 hours, and does not show intermediate lengths, as if an 'allowed time' was in effect here also. Conversely, making conditions more favourable increases the number of daughter cells without changing the time at which they appear. If *Chlorella* cultures are cooled to 4° C for 2 hours in the middle of a light period, no cell division takes place in that or the next light-dark cycle but occurs in the third dark period following the cold 'shock' at the usual time (Pirson, Lorenzen, and Koepper, 1959).

Edmunds (1966) found that synchronized cultures of *Euglena* removed from the synchronizing LD cycle to continuous dim light continued to show synchronous divisions for a very long time, at least 10 days. In these cultures too, some cells divided about every 24 hours, sometimes with a period of somewhat less than 24 hours. The proportion of cells which divide at one time may be of the order of 15%. If this result were due to the presence in the population of groups of cells synchronized with respect to each other but out of synchrony with other groups by 24 hours, 6–7 such groups must be postulated, certainly a more unlikely explanation than that cell division is under the control of a circadian timing mechanism.

Control over cell division by the circadian oscillator is probably not universal, since an LD cycle does not always synchronize division, but does universally entrain circadian rhythm. The common occurrence of generation times of about 24 hours, but quite independent of any light-controlled circadian rhythm may mimic these rhythms as soon as synchrony by any means is affected. Such periodicities do not persist in constant conditions for more than a short time however.

REFERENCES

Bernstein, E. (1960). Synchronous division in *Chlamydomonas moewusii*. *Science, N.Y.* **131**, 1528–1529.

Bernstein, E. (1964). Physiology of an obligate photoautotroph (*Chlamydomonas moewsuii*) I. Characteristics of synchronously and randomly reproducing cells and an hypothesis to explain their population curves. *J. Protozool.* **11**, 56–74.

Bruce, V. G. (1965). Cell division rhythms and the circadian clock. *In* Circadian Clocks, pp. 125–138 (J. Aschoff, ed.), North-Holland Publishing Co., Amsterdam.

Campbell, A. (1964). The theoretical basis of synchronization by shifts in environmental conditions. *In* Synchrony in Cell Division and Growth, pp. 469–484 (E. Zeuthen, ed.), Interscience Publishers, New York.

Cook, J. R. (1961). *Euglena gracilis* in synchronous division. II. Biosynthetic rates over the life cycle. *Biol. Bull. mar. biol. Lab Woods Hole.* **121**, 277–289.

Cook, J. R. (1966). The synthesis of cytoplasmic DNA in synchronized *Euglena*. *J. Cell Biol.* **29**, 369–373.

Cook, J. R., and M. Hess (1964). Sulfur-containing nucleotides associated with cell division in synchronized *Euglena gracilis*. *Biochim. biophys. Acta* **80**, 148–151.

Cook, J. R., and James, T. W. (1960). Light-induced division synchrony in *Euglena gracilis* var. *bacillaris*. *Expt. Cell Res.* **21**, 583—589.

von Denffer, D. (1949). Die planktonische Massenkultur pennatur Grunddiatomeen. *Arch. Mikrobiol.* **14**, 159–202.

Edmunds, L. N., Jr. (1964). Replication of DNA and cell division in synchronously dividing cultures of *Euglena gracilis*. *Science, N. Y.* **145**, 266–268.

Edmunds, L. N., Jr. (1965a). Studies on synchronously dividing cultures of *Euglena gracilis* Krebs (strain Z). I. Attainment and characterization of rhythmic cell division. *J. cell. comp. Physiol.* **66**, 147–158.

Edmunds, L. N., Jr. (1965b). Studies on synchronously dividing cultures of *Euglena gracilis* Krebs (strain Z). II Patterns of biosynthesis during the cell cycle. *J. cell. comp. Physiol.* **66**, 159–182.

Edmunds, L. N., Jr. (1966). Studies on synchronously dividing cultures of *Euglena gracilis* Krebs (strain Z). III. Circadian components of cell division. *J. cell. comp. Physiol.* **67**, 35–44.

Hastings, J. W., and Sweeney, B. M. (1964). Phased cell division in marine dinoflagellates. *In* Synchrony of Cell Division and Growth, pp. 307–321 (E. Zeuthen, ed.), Interscience Publishers, New York.

Hoogenhout, H. (1963). Synchronous cultures of algae. *Phycologia* **2**, 135–147.

Hotta, Y., and Stern, H. (1963). Molecular facets of mitotic regulation. I. Synthesis of thymidine kinase. *Proc. natn. Acad. Sci. U.S.A.* **49**, 648–654.

Hotta, Y. and Stern, H. (1963). Inhibition of protein synthesis during meiosis and its bearing on intracellular regulation. *J. Cell. Biol.* **16**, 259–280.

Hotta, Y. and Stern, H. (1966). Synthesis of DNA during meiosis. *Proc. natn. Acad. Sci. U.S.A.* **56**, 1184–1191.

Jørgensen, E. G. (1966). Photosynthetic activity during the life cycle of synchronous *Skeletonema* cells. *Physiologia. Pl.* **19**, 789–799.

Kates, J. R., and Jones, R. F. (1964). Variation in alanine dehydrogenase and glutamic dehydrogenase during the synchronous development of *Chlamydomonas*. *Biochim. biophys. Acta* **86**, 438–447.

Kates, J. R. and Jones, R. F. (1967). Periodic increases in enzyme activity in synchronized cultures of *Chlamydomonas reinhardti*. *Biochim. biophys. Acta* **145**, 153–158.

Leedale, G. F. (1959). Periodicity of mitosis and cell division in the Euglenineae. *Biol. Bull. mar. biol. Lab. Woods Hole.* **116**, 162–174.

Lewin, J. C., Reimann, B. E., Busby, W. F. and Volcani, B. E. (1966). Silica shell formation in synchronously dividing diatoms. *In* Cell Synchrony, pp. 169–188 (I. L. Cameron and G. M. Padilla, edrs) Academic Press, New York.

Lorenzen, H. (1959). Die photosynthetische Sauerstoffproduktion wachsender *Chlorella* bei langfristig intermittierender Belichtung. *Flora, Jena* **147**, 382–404.

Lorenzen, H. (1963). Temperatureinflüsse auf *Chlorella pyrenoidosa* unter besonderer Berucksichtigung der Zellentwicklung. *Flora, Jena* **153**, 554–592.

Lorenzen, H., and Schleif, J. (1966). Zur Bedeutung der kürzest möglichen Generationsdauer in Synchronkulturen von *Chlorella*. *Flora, Jena* **156**, 673–683.

Pirson, A., and Lorenzen, H. (1958). Ein endogener Zeitfaktor bei der Teilung von *Chlorella*. *Z. Bot.* **46**, 53–66.

Pirson, A., Lorenzen, H., and Koepper, A. (1959). A sensitive stage in synchronized cultures of *Chlorella*. *Pl. Physiol. Lancaster* **34**, 353–355.

Pirson, A. and Lorenzen, H. (1966). Synchronously dividing algae. *A. Rev. Pl. Physiol.* **17**, 439–458.

Pogo, A. O., and Arce, A. (1964). Synchronization of cell division in *Euglena gracilis* by heat shock. *Expt. Cell Res.* **36**, 390–397.

Prescott, D. M. (1964). The normal cell cycle. *In* Synchrony in Cell Division and Growth, pp. 71–97. (E. Zeuthen, ed.), Interscience Publishers, New York.

Rieth, A. (1939). Photoperiodizität bei zentrischen diatomeen. *Planta* **30**, 294–296.

Sachs, R. M., Bretz, C., and Lang, A. (1959). Cell division and gibberellic acid. *Expt. Cell Res.* **18**, 230–244.

Sorokin, C. (1960). Injury and recovery of photosynthesis in cells of successive development stages. The effects of light intensity. *Physiologia. Pl.* **13**, 687–700.

Sorokin, C. (1963). The capacity for organic synthesis in cells of successive development stages. *Arch. Mikrobiol.* **46**, 29–43.

Sorokin, C., and Krauss, R. W. (1961). Relative efficiency of photosynthesis in the course of cell development. *Biochim. biophys. Acta* **48**, 314–319.

Stern, H. (1961). Periodic induction of desoxyribonuclease activity in relation to the mitotic cycle. *J. biophys. biochem. Cytol.* **9**, 271–277.

Stern, H., and Hotta, Y. (1964). Regulated synthesis of RNA and protein in the control of cell division. *In* Meristems and Differentiation, pp. 59–72, Brookhaven Symposia in Biology, No. 16.

Sweeney, B. M., and Hastings, J. W. (1958). Rhythmic cell division in populations of *Gonyaulax polyedra*. *J. Protozool.* **5**, 217–224.

Sweeney, B. M., and Hastings, J. W. (1964). The use of the electronic cell counter in studies of synchronized cell division in dinoflagellates. *In* Synchrony in Cell Division and Growth, pp. 579–587 (E. Zeuthen, ed.), Interscience Publishers, New York.

Tamiya, H. (1967). Synchronous cultures of algae. *A. Rev. Pl. Physiol.* **17**, 1–26.

Volm, M. (1964). Die Tagesperiodik der Zellteilung von *Paramecium bursaria*. *Z. Physiol.* **48**, 157–180.

CHAPTER 8

The Mechanism for the Generation of Oscillations, Particularly those with a Circadian Period

All doubts with regard to the existence of rhythms in both plants and animals have now been dispelled by an enormous weight of evidence. Almost nothing is known however about the way in which rhythms are produced. At the heart of the matter is the generation of an oscillation from a constant source of energy. Of course, all oscillations may not be generated in the same way, but a thorough understanding of one such system would be a great step forward. Since in plants, at least, rhythms with a circadian period are best known at the present time and may underlie rhythms with lunar or annual periods, it seems fruitful to review the evidence which concerns the questions of the mechanism by which oscillations with a period of about 24 hours may be generated. Most of this evidence is still negative, and serves only to eliminate certain kinds of solutions to the puzzle.

Since the environment contains strong oscillations with a 24-hour period which arise from the rotation of the earth relative to the sun, the first possibility to resolve is whether this is the source of circadian oscillations for which we are seeking. The most obvious environmental periodicities are the diurnal changes in light and temperature. Since these factors can be controlled quite easily under laboratory conditions, it has been made clear for a long time that the circadian rhythms continue in the absence of fluctuations in light and temperature. Furthermore, the period of circadian rhythms in the free-running condition cannot be changed by previously keeping organisms in light or temperature cycles which are *not* 24 hours long. Such cycles cannot be learned. In fact, circadian rhythms can be induced in cells which have passed many generations in constant light and temperature simply by a single one-step difference in light intensity which contains no information with regard to cycle length. Therefore the daily changes in light and temperature are not directly responsible for circadian rhythms. Of course, they may have been highly instrumental in selecting a periodicity of this particular cycle length from among many arising spontaneously.

117

It now seems very unlikely that other environmental oscillations arising from the rotation of the earth relative either to the sun or to the stars, oscillations which cannot so easily be controlled in the laboratory as can light and temperature, could be responsible for the circadian rhythms either. Examples of variables which fall into this category are the strength of the earth's magnetic field and the amount or kind of radiation entering the earth's atmosphere from space. Other more subtle and elusive variables in this category can be imagined. Some of these factors, small magnetic fields for example, may be sensed by some organisms. However, because the oscillations in all such factors depend on the earth's rotation, their period must of necessity be exactly that of this rotation, i.e. 24 hours, and cannot vary at all with temperature. However, it is now very well established that the period of many circadian rhythms is not exactly that of the earth's rotation, and that the period does change when the level of either temperature or light intensity changes. Some of these changes in period are small but they are unmistakable. Given sufficient time, the maximum in a rhythm with a period not exactly 24 hours will scan through the day and night of solar time. This means that the phase of such a rhythm will eventually show every possible relationship to the position of the earth relative to the sun.

The principal untiring proponent of environmental periodicity as the source of circadian rhythms has been Professor Frank Brown. In order to understand his point of view, it is helpful to remember the course which his studies have followed. The subject of his early work on biological rhythms was the fiddler crab *Uca*. Brown discovered and documented a diurnal pattern of colour change in his crab which persists in darkness. This rhythm is unusually temperature independent, and under the environmental conditions which Brown and Webb (1948) used showed a period indistinguishable from 24 hours at temperatures between 6° and 26° C. However, light-dark cycles can reset this rhythm in the usual way, so that the crabs are always light-coloured in darkness. Brown was therefore quite naturally led to postulate that some environmental periodicity dependent upon the rotation of the earth was responsible, and that this periodicity did not determine the phase of the rhythm.

In subsequent experiments on potato tuber respiration, he showed that the rhythm in oxygen uptake, which is of very small magnitude and can only be detected by statistical means, is still present when barometric pressure is held constant. On this basis he ruled out this variable as the important one for rhythm generation. But he still persisted in his view that some other environmental factor could be invoked, encouraged by correlations of the amplitude of crab and potato rhythms with changes in barometric pressure, in the earth's magnetic field or with cosmic ray showers.

When it became clear that a number of rhythms show periods which are not exactly 24 hours and which can continue for a long time in constant conditions of light and temperature, and that, in fact, few other rhythms are as temperature insensitive as that in *Uca*, Brown recognized the difficulty that these observations posed for his theory of an environmental oscillator. His answer was to invoke a process which he termed 'auto-phasing'. In order to account for the periods which were observed in constant light and temperature and at the same time invoke as the source of rhythmicity an oscillator with a period strictly 24 hours long, one must postulate, either that a different phase adjustment is made in every cycle, or that the external oscillator imparts no information with regard to the phase of the biological oscillation that it drives and that this phase is shifted by a certain amount in each cycle, the changes being accumulated from cycle to cycle. The second alternative was favoured by Brown, and the mechanism for shifting phase in each cycle is what he calls 'auto-phasing' (Brown, 1965). Autophasing is postulated to arise from changes in the sensitivity to resetting which have been shown in many rhythms. As we have seen (Chapter 3) both the amount and the direction of the phase shift depends on the *time in the cycle* when cells are exposed to light in otherwise continuous darkness or continuous light regimes. Brown postulates that because of this changing sensitivity to light and darkness, a constant environment does not exert a constant effect on a rhythmic system. According to this view, this cycle in sensitivity to light constitutes the fundamental driven oscillation and hence must be completely temperature insensitive and show a period of exactly 24 hours. Its phase however will constantly shift relative to the external oscillator.

How does the theory of auto-phasing fit with what is known about the process of phase shifting by light and darkness? The problem that auto-phasing must explain is the presence of periods either shorter and longer than 24 hours in the same organisms depending on the ambient temperature. What would one predict from the phase response curves for the effect of a continuously applied stimulus, continuous light for example? Let us suppose that an organism is placed in continuous light at the start of a normal light period in its accustomed LD cycle. The phase response curve predicts that no shift in phase will occur during about the next 12 hours, which will appear to the organism as a normal light period of the LD cycle. However, if the light is bright enough to saturate the resetting mechanism, the phase will begin to be delayed as light extends into what was formerly the dark period. The evidence from plants like *Gonyaulax* is clear: the phase takes up its new position at once, and is now subject again to a new shift in phase position, another delay since the phase is still early night. Delay will thus follow delay and the middle of the night phase will never be attained. Thus rhythmicity would be predicted on these grounds to be abolished in bright light, a condition actually observed. If the light is less intense, smaller phase changes will be predicted. Since delays are smaller than advances even at saturating light intensity, perhaps a light intensity could be imagined where no delays at all would occur. At such an intensity of constant light, only advances would take place. However, as in the case of delays, the phase will be reset at once and then a second advance can occur. Due to the peculiarities of phase response curves, in particular the sharpness of the transition from delays to advances and the decrease in the magnitude of advances with time, perhaps rhythmicity could still be observed, now with a period appearing to be less than 24 hours, since a small advance had occurred in each cycle as postulated by autophasing. However, periods longer than 24 hours could never occur by this mechanism. Furthermore, since oscillation in the environment is not affected by temperature, only one apparent period length shorter than 24 hours could be produced in any one organism by autophasing. However, many different periods, both shorter or longer than 24 hours, have been observed in *Gonyaulax, Oedogonium, Euglena* and other organisms subjected to different constant temperature levels. The phase response curve in fact can be considered just another manifestation of innate rhythmicity which itself can show periods other than 24 hours. Hence autophasing cannot be used to justify external environmental variables as the source of the oscillation cuing the circadian rhythms.

More direct evidence also makes dubious the invoking of celestial variables. No effect of any kind on rhythms has been shown in experiments where either pulses or constant levels of magnetic flux or high energy radiation have been used. A consideration of the high variability of these parameters which is superimposed on the cyclic component in nature poses a real problem in detection for any organism, were it to utilize such a source of information on period or phase.

A further objection to the theory of an external oscillator is the phase lability required. In physical oscillations, such freedom is not encountered, while mechanisms exist for producing any given phase relationship between driving and driven oscillator, this relationship is fixed, once established.

What mechanism, then, can be responsible for the oscillation which we observe as circadian rhythms? The oscillator must be endogenous if it is not of environmental origin. This conclusion is supported by demonstrations that replacing hydrogen with deuterium in the components of living cells lengthens the period of their rhythmicity. Bruce and Pittendrigh (1960) conducted such an experiment with *Euglena*. The period of the phototactic rhythm in this organism is $23\frac{1}{2}$ to $23\frac{3}{4}$ hours in water, but after the cells have been grown in D_2O for some time, the period is lengthened to $26\frac{1}{2}$ to 28 hours. A 3-day exposure to deuterium oxide also shifts the phase of the *Euglena* rhythm. Bünning and Baltes conducted similar experiments with *Phaseolus*. Short exposures to pure D_2O at midday caused about a 12-hour reset in the leaf movement rhythm. When grown in 50% D_2O, the bean rhythm also showed a longer period than in water. The substitution of a heavier atom for hydrogen might be expected to slow an internal physical or chemical oscillator, but would hardly affect the period of an environmental periodicity.

Oscillations can arise in cell-free chemical or biochemical systems. As we have seen (Chapter 5), extracts of yeast containing the enzymes concerned with glycolysis oscillate on the addition of substrates (Betz and Chance, 1965a, b). The period of these oscillations is very short at room temperatures, but may be as long as 1·5 hours if the temperature is lowered. The oscillations are thought to be the result of positive feedback effects. Their existence provides a precedent for the view that a similar mechanism could give rise to oscillations with a much longer period. These oscillating biochemical events in yeast and in extracts from yeast are temperature-dependent however, so that they cannot serve as a model for the circadian system without further assumptions. Short period oscillations have also been observed in H^+ and K^+ fluxes from isolated mitochondria in the presence of valinomycin which increases the permeability of membranes to K^+ (Chance *et al.* 1964, Chance and Yoshicka, 1966). These oscillations too are temperature sensitive, and damp out fairly rapidly. They do however provide another example of an *in vitro* system which can oscillate.

Electronic models which yield oscillations with the general properties of circadian rhythms, particularly the typical phase response curves to light and temperature pulses, have been constructed by Sage *et al.* (1962, 1963). The oscillations generated in this way have very short periods, but they serve to demonstrate that it is indeed possible to derive oscillations with phase determination like that of circadian rhythms from circuits incorporating feedback control techniques. The problem of the origin of relative temperature independence is not amenable to such methods, however. The effects of both constant and fluctuating temperature on circadian clocks can be represented in mathematical models, as shown recently by Pavlidis *et al.* (1968).

E

The construction of meaningful models for circadian oscillators is hampered by the fact that there is no way at present to measure the basic oscillator directly. Thus it is not possible to draw any conclusion concerning the shape of the oscillation which it generates. Since the rhythms that we are able to measure show diverse shapes, even in the same organism (compare the forms of the rhythms in luminescence and cell division in *Gonyaulax*, for example), they probably do not reflect the oscillator in this respect. While the leaf movement rhythm in *Phaseolus* at low temperature appears to show the characteristics of a relaxation oscillation (Bünning, 1960), many overt rhythms more closely resemble a sine wave in form. It is thus not possible to make a choice with confidence even between these two basic wave forms for the driving oscillator. It is not surprising then that models have been only of limited usefulness in understanding circadian oscillations.

What do we know about where to look for the oscillator responsible for circadian rhythmicity? The evidence at hand is almost entirely of a negative nature, eliminating some candidates without committing any specific site. In the first place, it is clear that the oscillator does not require the interaction of several cells. Isolated tissues from higher plants, for example carrot tissue and the mesophyll of *Bryophyllum* grown in culture, exhibit rhythmicity. Similarly, in animal tissue, Strumwasser (1965) has demonstrated a rhythm in spontaneous firing of a single neuron in an isolated ganglion from the sea hare, *Aplysia*. Furthermore, populations of many single-celled organisms are rhythmic in culture. Interactions between cells via changes in the culture medium are ruled out as the source of oscillation, since the photosynthesis of the unicellular dinoflagellate *Gonyaulax* has been measured from a single cell in isolation and shown still to be rhythmic (Sweeney, 1960). Several dinoflagellates have been isolated in capillaries and continue their rhythmic cell division under such circumstances (Sweeney, 1959). *Acetabularia*, too, is rhythmic when a single cell is studied (Schweiger and Berger 1964). Furthermore, extracts of cells in one phase of the circadian cycle do not alter the rhythm of cells of the opposite phase when small amounts are added to the medium (Sweeney, unpublished). Mixing *Gonyaulax* cultures which are out of phase does not give any evidence of interaction (Hastings and Sweeney, 1958). In this experiment, the luminescence was measured in the absence of external light and temperature cues, and the curve was that expected from the addition of the component cultures. Interaction between different organelles within a single cell is not ruled out by any of these experiments.

A number of investigators have considered the possibility that the oscillator could be identified with a particular cellular organelle. To implicate a specific organelle as the site of the oscillator it would be necessary to show that rhythms in other unaffected organelles or in the cell as a whole ceased in its absence. Although cell division can be ruled out as the circadian oscillator as discussed in Chapter 7, still the nucleus is an attractive candidate for the location of a circadian clock because of its central role in protein synthesis. *Acetabularia* offers a unique opportunity to test this possibility, because it is uninucleate and can survive enucleation for long periods with apparently unimpaired protein synthesis and photosynthesis. Fortunately, too, a circadian rhythm in photosynthesis can be found in a number of species of *Acetabularia*. However, when a test of the hypothesis that the nucleus is the site of the circadian oscillator in *Acetabularia* was made, the rhythm in photosynthesis was found to persist equally well in nucleated and enucleated cells, even under conditions of constant light and temperature. Thus a nuclear oscillator cannot be driving the photosynthetic rhythm in *Acetabularia*. Schweiger *et al.* conducted very interesting experiments in which the nucleus of an *Acetabularia* in one phase was transplanted into an enucleated cell of the opposite phase, or the base of the cell, which contains the nucleus, was grafted on to a top of another cell of the opposite phase. In both cases, the phase of the resultant photosynthetic rhythm was that of the plant donating the nucleus rather than that of the plant actually supplying the photosynthetic apparatus. The nucleus of *Acetabularia* then must also house some kind of circadian rhythm, one which can control cytoplastic activities including the photosynthetic rhythm. The interaction of these apparently separate circadian oscillators is a most interesting phenomenon, and deserves further study.

Bünning and his co-workers have detected two changes in the nucleus correlated with time of day. In *Allium cepa*, Bünning and Schöne-Schneiderhöhn (1957) report that the nucleus is larger at noon and smaller at night and that these changes can still be detected on the second day in LL. In sections of the leaf cells of *Elodea canadensis* examined with the aid of the electron microscope, layers of the nuclear membrane appear to be separated at night more widely than during the day. However, cells from which the sections were made were not in constant conditions, so that the circadian nature of these changes is not established.

To examine the possibility of circadian morphological changes, cells of *Gonyaulax polyedra* were fixed at different times of day in light-dark cycles and in constant light of low enough intensity so that the rhythm in luminescence was observed (Bouck and Sweeney, unpublished). Examination of thin sections under the electron microscope (see frontispiece) failed to show any differences in structure which could be correlated with the time in the cycle when the cells were fixed, although nucleus, nuclear membrane, Golgi bodies, mitochondria, plastids and cell membranes were well preserved.

Vanden Driessche (1966) has reported that the chloroplasts of nucleated and enucleated *Acetabularia* show a circadian rhythm in shape, becoming more elongated in the middle of the day phase and more spherical at night. However, since these changes have not been demonstrated in isolated chloroplasts, it is not possible to conclude anything from them concerning the location of the oscillator responsible. They may well simply be another manifestation of the rhythm in the rate of photosynthesis.

Is it possible to localize the basic oscillation in any specific biochemical process? This is a difficult question to investigate, because of the prevalence of rhythms, many of which obviously depend on biochemical changes, for example those in photosynthesis, luminescence and carbon dioxide fixation. There is no easy way to distinguish a causative oscillation from the many rhythmic processes, both known and possibly not yet described, which this oscillator theoretically controls. The only unique properties of the oscillation as distinct from these other rhythmically-varying biochemical processes are period and phase, both of which it imparts to the processes which it controls. Therefore a reaction sequence which belongs to the oscillator might be identified by the fact that changing it in a specific way alters either the phase or the period of the rhythms we can observe. Abolishing rhythmicity by specific inhibitors does *not* implicate the process for which the inhibitor is specific as a part of the clock, since the inhibition may simply have affected the process which the clock controls. The oscillator may be operating as usual in the presence of the inhibitor, but can no longer be measured. Furthermore, there must be many links responsible for the transduction of information from the oscillator to the overt rhythmic processes serving as indicators. One of these coupling reactions may have been affected by the inhibitor in question. The evidence at hand suggests that this transducing sequence is a one-way street: information cannot flow back to the oscillator from the processes under its control to alter period and phase. This can be seen in experiments with *Gonyaulax* in which photosynthesis is partially inhibited by the specific inhibitor, dichlorophenyl dimethyl urea, at very low concentrations. Although sufficient DCMU is present to bring about 50% inhibition of photosynthesis, the rhythm in carbon assimilation continues with the same period as before its addition. Furthermore, 3-hour exposures to DCMU which completely inhibit photosynthesis, do not change the phase of the rhythm observed on removal of the inhibitor, and the luminescent rhythm is similarly unaffected. Discharge of luminescence on stimulation, which must alter the state or the amount of luciferin, likewise shows no effect on the phase of the rhythm in luminescence (Hastings and Sweeney, 1958). Inhibitors of cell division do not change the phase or the period of other rhythms in *Gonyaulax*. It is not yet certain that all the rhythms in *Gony aulax* are under the control of a single oscillator. However, during long intervals in constant conditions, the rhythms of luminescence and cell division retain the same phase relationship. This indicates that they run at exactly the same frequency. Furthermore, their temperature-dependence and resetting appear to be alike. A single oscillator is thus the simplest assumption in the absence of the necessity for more than one.

The absence of feedback from photosynthesis, luminescence and cell division to the timing oscillation, changes in the rates of these processes having no effect on the period or phase of rhythms at least in *Gonyaulax*, rules out these processes or any of their component biochemistry as the source of oscillation. The clock must be located elsewhere, exerting its control via transducing mechanisms without feedback.

A long series of experiments has been carried out in which plants or animals which show circadian rhythms have been treated with various biologically active substances, and changes in either period or phase have been sought, with the hope of detecting the oscillator in this way. In general, they have been conspicuously unsuccessful. When shortening or lengthening of the period has been chosen as the indicator of effectiveness, substances have been applied and the organism has been kept in constant conditions. This restricts the substances which can be used to those which do not have highly toxic effects during long exposures, because the rhythm in question must be observable for much longer than a single cycle to provide accurate measurements of period, the more so since the expected changes might be quite small.

Bühnemann's experiments with the sporulation rhythm of *Oedogonium*, were among the first in which the effect of inhibitors on rhythms was explored. He found that cyanide, azide, dinitrophenol, arsenate and iodoacetate, at concentrations which depressed sporulation markedly, had no effect on the period of rhythmicity. The rhythm could still be observed even when almost no spores were being produced. Respiration thus is not the oscillator. In fact, very little respiratory energy is needed to sustain oscillations, although completely anaerobic conditions usually stop them. Bühnemann tested other substances including copper sulphate, cocaine, indole acetic acid, adenosine triphosphate and riboflavin, none of which had any effect on the rhythm in sporulation in *Oedogonium*.

Many experiments since Bühnemann's have served to underline the surprising insensitivity of rhythmicity to the presence of various chemicals. Some changes in leaf movement in *Phaseolus* plants cut off and

dipped in alcohol, cochicine, or urethane solutions were reported by
Bünning (1960), Keller (1960) and Bünning and Baltes (1963) (FIG. 8-1).
The leaf movements appeared disturbed at first but the original rhythm
was sometimes restored, making it difficult to decide whether the basic
oscillation or just the process of leaf movement had been affected.
In connection with the pulse experiments described below, Hastings
measured the period of the glow rhythm in *Gonyaulax* in LL in the

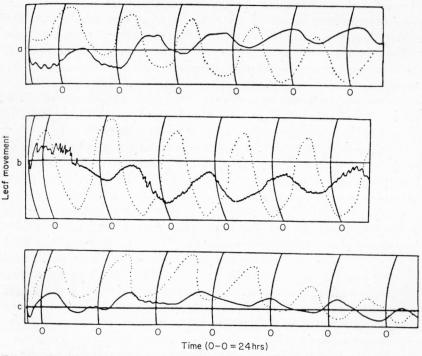

FIG. 8-1. The leaf movement rhythm of the primary leaves of *Phaseolus multiflorus* in
continuous light (400–900 lux) at 20° C, in the presence of 2% ethanol (curve **a**); 5%
methanol (curve b); and 0·01% Victoria blue (curve c); The curves for untreated plants
are shown dotted. (Keller, 1960.)

presence of a number of substances which stimulate or inhibit cellular processes, but he could obtain no evidence for any change in
period by any of them. Experiments in which I measured the rhythm
in stimulated luminescence in the presence of chloramphenicol and colchicine over long periods of time in continuous light did give evidence
of small changes in period, an hour's lengthening in the presence of
colchicine, 10^{-3}M, and a shortening of the period by 2 hours in chloramphenicol 2×10^{-4}M, the most effective concentrations. The glow rhythm
did not show any effect of chloramphenicol (10^{-4}M) in the experiments

of Hastings. This discrepancy is still to be resolved. In any case the period changes were very small. Substances which do not change the period of *Gonyaulax* rhythms include cobaltous ion, indole acetic acid, kinetin, gibberellic acid, arsenite, arsenate, PCMB, dinitrophenol, cyanide, urethane, chloropromazine, DCMU and CMU. Puromycin mitomycin C, novobiocin, terramycin, bacitracin, EDTA, sodium azide, $FeCl_3$, HgCl, $CaCl_2$, KCl, $AgNO_3$, all of necessity assayed at concentrations which allowed rhythmicity to be observed.

Equally frustrating were the results of experiments where substances were applied for short times at different parts of the circadian cycle and phase changes were sought. A long series of such experiments were

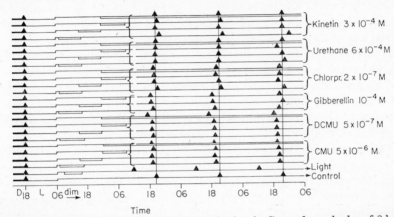

FIG. 8-2. The effect on the phase of the glow rhythm in *Gonyaulax polyedra* of 8-hour exposures to different biologically active substances. Each line represents a single experiment in which a substance was added to a cell suspension during the time indicated by the underlining, and then removed by centrifugation. The time of maximum glow on the subsequent days in dim continuous light is indicated by a black triangle. The substances used are indicated at the right of the figure. Exposure to bright light instead of chemicals served as a control. (Hastings, 1960. *Cold Spring Harb. Symp. quant. Biol.* **25**, 131–145).

carried out by Hastings with the *Gonyaulax* glow rhythm (FIG. 8-2). Sometimes substances appeared to alter phase but when the treatments were repeated, different or no phase changes were obtained. Of a large number of substances tested, not a single one gave phase changes at all comparable to those which short exposures to light produced. Among those substances which brought about variable amounts of phase change were arsenite and PCMB. Hastings concludes that the accuracy of timing may be reduced by such substances. No evidence for the biochemical nature of the oscillator was forthcoming, however.

The discovery of the steps by which messenger RNA is synthesized on a DNA template and subsequently directs the synthesis of protein led to the suggestion that this sequence might be responsible for the generation of oscillations of a circadian period. While the amount of incorporation of ^{32}P could be shown to vary cyclically in some organisms such as *Phaseolus* (Rückebeil, 1961) this was not the general rule. No differences were detected in *Gonyaulax*, for example (Hastings, 1960). However, perhaps only some specific RNA is important in oscillations and this may be masked by the larger non-cyclic fraction of RNA synthesis. Rhythmic incorporation of labelled amino acids into protein in *Euglena* under conditions where the rhythm of phototaxis persists has recently been demonstrated by Feldman (1967). The evidence that steps in protein synthesis constitute the basic oscillator rests on shaky ground however. Evidence in favour of this hypothesis concerns first the effectiveness of ultraviolet light of about the wavelength absorbed by nucleic acids in resetting the mating rhythm in *Paramecium* (Ehret, 1960) and the luminescent rhythm in *Gonyaulax* (Sweeney, 1963). Then, studies in which *Gonyaulax* cells were exposed to actinomycin D at very low concentrations and the glow rhythm recorded (Karakashian and Hastings, 1962, 1963), showed that rhythmicity disappeared in the presence of this specific inhibitor of mRNA synthesis, leaving intact a small but constant glow (FIG. 8-3). However, disappearance of a rhythm in itself is not very compelling evidence for specific effects on the clock, since coupling mechanisms rather than the basic oscillator may have been inhibited. Moreover, the small amount of luminescence retained in the presence of actinomycin D sometimes seemed to show suspicious fluctuations.

Other rhythms in *Gonyaulax* do not react as does the glow rhythm to the presence of actinomycin D. The photosynthetic rhythm is not at all changed by a concentration of actinomycin D which eliminates the glow rhythm, in fact the data are open to question that even concentrations ten times higher are effective. Furthermore, if the oscillator consists of protein synthesis on an RNA template, why does not chloramphenicol stop rhythmicity or at least show affects on period or phase? It penetrates *Gonyaulax* cells since cell division is inhibited in its presence.

Are rhythms in other organisms sensitive to actinomycin D? Strumwasser has shown that the rhythmic firing of a single neuron of *Aplysia* can be induced to start earlier by the addition of actinomycin D before the spike is expected. The next firing also occurs earlier, 24 hours after the first, so that the rhythm appears to have been reset by the addition. MacDowell (1964) has reported that exudation of sap from detopped tobacco plants, known for a long time to be rhythmic, shows a lengthened period in the presence of a low concentration of actinomycin D (0·25 μg per ml. in the nutrient solution surrounding the roots). However, only a single experiment with one concentration of this inhibitor was done, and so it is impossible to assess the repeatability of this observation and whether or not the lengthening of the period is concentration-dependent.

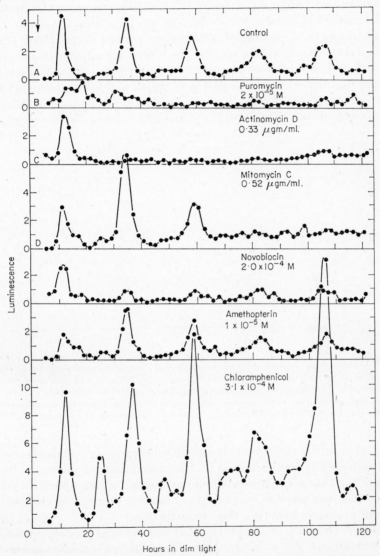

FIG. 8-3. Recordings of the glow emitted from *Gonyaulux polyedra* during exposure to substances which affect the synthesis of macromolecules. The cell suspensions were kept in continuous light except while recording was in progress. Substances as indicated were added at the time indicated by the arrow (Karakashian and Hastings, 1963.)

If RNA synthesis is the site of the generation of oscillation in *Acetabularia*, the DNA of the nucleus cannot be acting as the only template, since enucleate cells still maintain a normal photosynthetic rhythm. Of course, the nucleus is not the only organelle to contain a DNA complement and to be able to synthesize RNA as a consequence. In *Acetabularia*, isolated chloroplasts have been shown to contain DNA and to be able to direct the formation of RNA (Schweiger and Berger 1964). Is this RNA synthesis generating oscillations with a circadian period? When inhibitors of either RNA and protein synthesis were present in sea water containing *Acetabularia crenulata* cells which had been enucleated and the photosynthetic rhythm was measured in continuous light and constant temperature (FIG. 8-4), there was no evidence for any change in

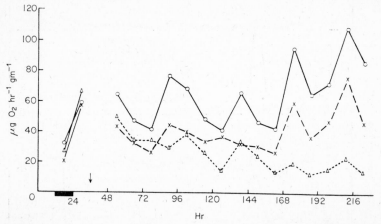

FIG. 8-4. The rhythm in photosynthesis in enucleated *Acetabularia crenulata* in continuous light (500 lux) at 17° C in the presence of actinomycin D, 20 μg ml^{-1} (×); chloramphenicol 250 μg ml^{-1} (△); control (○). Inhibitors were first added at the time indicated by the arrow and again after every subsequent oxygen determination. Photosynthesis was measured as net oxygen evolved during 1 hr. at 5000 lux, and 17° C. (Sweeney, Tuffli and Rubin, 1967.)

the period of the rhythm, although the incorporation of radioactivity into amino acids was decreased by the presence of the inhibitors by at least 50%. That actinomycin D, puromycin and chloramphenicol penetrated into the cell was obvious also from the decrease in the rate of oxygen evolution following their introduction.

Vanden Driessche has reported that the rhythms of photosynthesis and chloroplast shape in *Acetabularia mediterranea* are inhibited by actinomycin D (2·7 μg per ml.) when the nucleus is present but not in its absence. Her measurements were spaced 12 hours apart and her data recorded as the ratio of the day rate of photosynthesis to that at night when the cells were in continuous light. In the presence of actinomycin these ratios approached 1·0 in cells with nuclei in the presence of actinomycin. However, it is not possible to be sure that a rhythm of small amplitude did not remain. In continuous light, too, the period is likely to be different from 24 hours and so the measurements may not have been taken exactly at the peak and trough of the oscillation. A reduction in the amplitude of the rhythm is not significant when effects on the underlying oscillator are to be considered, since such observations may well represent only the decrease in overall photosynthesis without any underlying effect on the timing of the cycles.

All in all, it seems premature to build theories in which the generation of circadian oscillations is attributed to the transcription of the genetic message, as Goodwin (1963) and Ehret and Trucco (1967) have done, particularly since no dramatic changes in the incorporation of radioactive DNA or RNA precursors into *Paramecium* correlated with cycle phase can be detected. Since protein synthesis as a whole is not rhythmic, the whole genome cannot be involved. Specialized portions of the chromosome, the chronons postulated by Ehret and Trucco, would still be expected to be sensitive to actinomycin D. The failure of some rhythmic systems to be affected by this substance thus requires explanation. Furthermore, a mechanism invoking *m*RNA and subsequently protein synthesis should always be very sensitive to inhibitors of protein synthesis like chloramphenicol, puromycin, and cycloheximide. Only in *Euglena* have changes in period in the presence of cycloheximide been observed. To be sure, these changes are concentration-dependent (FIG. 8-5), but the lengthening of the period in the presence of cycloheximide is small compared to the inhibition of incorporation of label from radioactive amino acids into protein measured in similar *Euglena* cells. Again special assumptions regarding a specific small portion of the total protein synthesis, less sensitive to inhibition, must be made.

Then, too, the existence in *Euglena* of a rhythm in incorporation of phenylalanine in continuous darkness does not necessarily mean that steps in protein synthesis are oscillator components because protein synthesis may be under the control of another rhythmic variable. In fact an underlying rhythm in the availability of high energy compounds is suggested by the observation that when light is present during incorporation of phenylalanine, protein synthesis no longer varies with time in the cycle, but is always uniformly high. Furthermore, the chronon theory, including the mechanism by which a new round of transcription is initiated, does not account for the phase response curves so characteristic of circadian rhythms in general.

Inhibitor studies have thus left us still in doubt concerning the generation of oscillations. How then can we hope to approach the problem of the mechanism which must underlie the production of the circadian rhythmicity? We have at our disposal only processes which show the presence of the oscillator but are not themselves a part of it. In order to control these processes however, the oscillator, whatever its nature, must be able to transmit information with regard to phase and period to the processes which it controls. We have already discussed the evidence which leads to the conclusion that this transduction takes place in only one direction, from oscillator to overt rhythm. However, if the nature of this coupling was understood, perhaps conclusions concerning the nature of the oscillator would be possible. Pursuing this line of

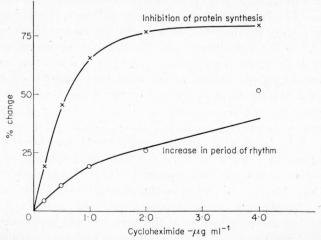

FIG. 8-5. The effect of cycloheximide on the length of the period of the rhythm in phototaxis in *Euglena gracilis* Z strain in DD (○) compared with the effect of this inhibitor on protein synthesis as measured by the incorporation of ^{14}C phenylalanine under the same conditions (X). The period in untreated *Euglena* is 23·7 hours. (Data replotted from Feldman, 1967.)

reasoning, I have been examining photosynthesis in *Gonyaulax* to find where transduction is taking place and what kind of process it is. The rhythm in photosynthesis in *Gonyaulax* offers a unique opportunity, since only the capacity, not the actual rate of photosynthesis is changing, when cells are maintained in constant light of about 500–1000 lux. The constant rate of photosynthesis under these conditions assures that fluctuations in the amounts of intermediates resulting from different rates of production and utilization will not confuse the interpretation. Under these conditions, not all the partial processes making up photosynthesis as a whole are rhythmic. While the capacity for overall photosynthesis is changing, that for carrying out the Hill reaction is not. The pigment content is constant as is the sensitivity to inhibition by DCMU and by carbonyl cyanide *m*-chlorophenylhydrazone, an inhibitor of photosynthetic phosphorylation. However, when crude cell extracts are prepared from *Gonyaulax* at different phases of the rhythm, the activity of the ribulose diphosphate carboxylase in these extracts does vary, and its activity faithfully reflects the photosynthetic capacity of whole cells, as long as the concentration of bicarbonate in the assay mixture is low. When assays are made of the activity of this enzyme at high bicarbonate concentrations however, extracts made at the time when the photosynthetic capacity of the cells is low are as active as those made from cells with a high photosynthetic capacity. In living cells also, increasing the bicarbonate concentration reduces the difference in the photosynthetic capacity as a function of cycle phase, although high enough bicarbonate concentrations cannot apparently be attained inside the cells to abolish rhythmicity completely. This result indicates that synthesis and destruction of the enzyme are not taking place and hence the oscillator does not control photosynthesis via protein synthesis. Changes which alter activity rather than amount of enzyme are indicated.

A similar analysis of the rhythm in stimulated luminescence leads to interesting speculation. If the luciferase and the luciferin are extracted from *Gonyaulax* at different times in the rhythmic cycle under constant conditions, higher activities of both are found during the night phase when luminescence is brighter (Bode, 1961, Bode *et al.*, 1963). However, while the differences in activity are at best about 5 times, the luminescence in cells is about 50 times brighter at night than during the day. This large discrepancy can be traced to another factor, namely changes in the stimulatory machinery. In *Gonyaulax*, luminescence is usually stimulated mechanically, in response to a sharp shake or other disturbance. At night a flash of light occupying about a tenth of a second is produced. If the cells are in the day phase and in light, much more vigorous shaking is necessary to produce any bioluminescence and the kinetics of the flash are different. A smaller difference in flash kinetics between day and night phase cells is seem in continuous darkness. The generation of a flash is probably preceded by the initiation of an action potential. At least, the importance of such action potentials is now well established in another dinoflagellate, *Noctiluca*, by the painstaking studies of Eckert. The generation of action potentials depend on sudden changes in membrane permeability to ions. In *Noctiluca*, the membrane in question appears to be common to the whole cell, since action potentials generated at one point on the surface are propagated around the cell with a predictable time delay. If the large rhythmic component which concerns the stimulation of bioluminescence derives from changes in the properties of this membrane, then a mechanism by which control over a large number of diverse processes could be exerted by a central oscillator could be imagined.

Logic demands that any general solution to the problem of the generation of oscillations in so many and such diverse cells must invoke a universal and hence a fundamental process in cellular physiology and biochemistry. This has been the attraction of theories invoking transcription of the genome. Membrane behaviour is another common denominator of cellular organization. Perhaps as membranes become better understood, the solution of the problem of the mechanism of the cellular clocks will become apparent.

REFERENCES

Betz, A., and Chance, B. (1965). Influence of inhibitors and temperature on the oscillation of reduced pyridine nucleotides in yeast cells. *Archs. Biochem. Biophys.* **109,** 579–584.
Betz, A., and Chance, B. (1965a). Phase relationships of glycolytic intermediates in yeast cells with oscillatory metabolic control. *Archs. Biochem. Biophys.* **109,** 585–594.
Bode, V. C. (1961). The bioluminescent reaction in *Gonyaulax polyedra*: a system for studying the biochemical aspects of diurnal rhythms. Ph.D. Thesis, University of Illinois, Urbana.
Bode, V. C., DeSa, R., and Hastings, J. W. (1963). Daily rhythm of luciferin activity in *Gonyaulax polyedra*. *Science, N.Y.* **141,** 913–915.
Brown, F. A., Jr. (1965). A unified theory for biological rhythms: rhythmic duplicity and the genesis of 'circa' periodisms. *In* Circadian Clocks, pp. 231–261 (J. Aschoff, ed.), North-Holland Publishing Co., Amsterdam.
Brown, F. A., Jr., and Webb, H. M. (1948). Temperature relations of an endogenous daily rhythmicity in the fiddler crab, *Uca*. *Physiol. Zoöl.* **21,** 371–381.
Bruce, V. G., and Pittendrigh, C. S. (1960). An effect of heavy water on the phase and period of the circadian rhythm in *Euglena*. *J. cell. comp. Physiol.* **56,** 25–31.
Bühnemann, F. (1955). Das endodiurnale System der *Oedogonium* zelle. II. Der Einfluss von Stoffwechselgiften und anderen Wirkstoffen. *Biol. Zbl.* **74,** 691–705.
Bünning, E. (1960). Opening address: biological clocks. *Cold Spring Harb. Symp. quant. Biol.* **25,** 1–9.
Bünning, E. (1964). The Physiological Clock. Springer-Verlag, Berlin.
Bünning, E., and Baltes, J. (1963). Zur Wirkung von schwerem Wasser auf die enfogene Tagesrhythmik. *Naturwissenschaften,* **50,** 622.
Bünning, E., and Könitz, W. (1962). Tagesperiodische Änderungen in den Kernmembranen, *Naturwissenschaften* **49,** 20.
Bünning, E., and Schöne-Schneiderhöhn, G. (1957). Die Bedeutung der Zellkerne in Mechanismus der endogenen Tagesrhythmik. *Planta* **48,** 459–467.
Chance, B., Schoener, B., and Elsaesser, S. (1964). Control of the waveform of oscillations of the reduced pyridine nucleotide level in a cell-free extract. *Proc. natn. Acad. Sci. U.S.A.* **52,** 337–341.

Chance, B., and Yoshioka, T. (1966). Sustained oscillations of ionic constituents of mitochondria. *Archs. Biochem. Biophys.* **117**, 451–465.

Eckert, R. (1965). Bioelectric control of bioluminescence in the dinoflagellate *Noctiluca. Science N.Y.* **147**, 1140–1145.

Ehret, C. F. (1960). Action spectra and nucleic acid metabolism in circadian rhythms at the cellular level. *Cold Spring Harb. Symp. quant. Biol.* **25**, 149–158.

Ehret, C. F., and Trucco, E. (1967). Molecular models for the circadian clock. I. The Chronon concept. *J. theor. Biol.* **14**, 240–262.

Enderle, W. (1951). Tagesperiodische Washstums- und Turgorschwankungen an Gewebekulturen. *Planta* **39**, 570–588.

Feldman, J. F. (1967). Lengthening the period of a biological clock in *Euglena* by cycloheximide, an inhibitor of protein synthesis. *Proc. natn. Acad. Sci. U.S.A.* **57**, 1080–1087.

Feldman, J. F. (1968). Circadian rhythmicity in amino acid incorporation in *Euglena gracilis. Science, N.Y.* **160**, 1454–1456.

Goodwin, B. C. (1963). Temporal Organization in Cells, Academic Press, London and New York. 163 pp.

Hastings, J. W. (1960). Biochemical aspects of rhythms: phase shifting by chemicals. *Cold Spring Harb. Symp. quant. Biol.* **25**, 131–143.

Hastings, J. W., and Keynan, A. (1965). Molecular aspects of circadian systems. *In* Circadian Clocks, pp. 167–182 (J. Aschoff, ed.) North-Holland Publishing Co., Amsterdam.

Hastings, J. W., and Sweeney, B. M. (1958). A persistent diurnal rhythm of luminescence in *Gonyaulax polyedra. Biol. Bull. mar. biol. Lab., Woods Hole* **115**, 440–458.

Karakashian, M. W., and Hastings, J. W. (1962). The inhibition of a biological clock by actinomycin D. *Proc. natn. Acad. Sci. U.S.A.* **48**, 2130–2137.

Karakashian, M. W., and Hastings, J. W. (1963). The effect of inhibitors of macromolecular synthesis upon the persistent rhythm of luminescence in *Gonyaulax. J. gen. Physiol.* **47**, 1–12.

Keller, S. (1960). Über die Wirkung chemischer Faktoren auf die tagesperiodischen Blattbewegungen von *Phaseolus multiflorus. Z. Bot.* **48**, 32–57.

MacDowell, F. D. H. (1964). Reversible effects of chemical treatments on the rhythmic exudation of sap by tobacco roots. *Can. J. Bot.* **42**, 115-122.

Pavlidis, T., Zimmermann W. F., and Osborn, J. (1968). A mathematical model for the temperature effects on circadian rhythms. *J. theor. Biol.* **18**, 210–221.

Richter, G., (1963). Die Tagesperiodik der Photosynthese bei *Acetabularia* und ihre Abhangigkeit von Kernaktivitat, RNS- und Protein-synthese. *Z. Naturf.* **18**, 1085–1089.

Rückebeil, A. (1961). Untersuchungen zur Tagesperiodizität der [32]P Einlagerung in die Verbindungen der Nukleinsäurefraktionen. *Z. Bot.* **49**, 1–22.

Sage, A. P., Jnr., Cockrum, E. L., Justice, K. E., and Melsa, J. L. (1962). Study and research on electronic simulation of the biological clock. Technical Documentary Report No. ASD-TDR-62-191, pp. 1–81.

Sage, A. P., Jr., Justice, K. E., and Melsa, J. L. (1963). Study and research on electronic simulation of the biological clock. II. Technical Documentary Report No. ASD-TDR-63-136, pp. 1–66.

Schweiger, E., Wallraff, H. G., and Schweiger, H. G. (1964). Über tagesperiodische Schwankungen der Sauerstoffebilanz kernaltiger und kernloser *Acetabularia. mediterranea. Z. Naturf.* **19b**, 499–505.

Schweiger, E., Wallraff, H. G., and Schweiger, H. G. (1964). Endogenous circadian rhythm in cytoplasm of *Acetabularia*: influence of the nucleus. *Science, N.Y.* **146**, 658–659.

Schweiger, H. G., and Berger, S. (1964). The DNA dependent RNA synthesis in chloroplasts of *Acetabularia*. *Biochim. biophys. Acta* **87**, 533–535.

Strumwasser, F. (1965). The demonstration and manipulation of a circadian rhythm in a single neuron. *In* Circadian Clocks, pp. 442–462. (J. Aschoff, ed.), North-Holland Publishing Co., Amsterdam.

Sweeney, B. M. (1959). Endogenous diurnal rhythms in marine dinoflagellates. International Oceanographic Congress Preprints pp. 204–207 (M. Sears, ed.), A.A.A.S., Washington.

Sweeney, B. M. (1960). The photosynthetic rhythm in single cells of *Gonyaulax polyedra. Cold Spring Harb. Symp. quant. Biol.* **25**, 145–148.

Sweeney, B. M. (1963). Resetting the biological clock in *Gonyaulax* with ultra-violet light. *Pl. Physiol. Lancaster,* **38**, 704–708.

Sweeney, B. M. (1965). Rhythmicity in the biochemistry of photosynthesis in *Gonyaulax. In* Circadian Clocks, pp. 190–194 (J. Aschoff, ed.) North-Holland Publishing Co., Amsterdam.

Sweeney, B. M., and Haxo, F. T. (1961). Persistence of a photosynthetic rhythm in enucleated *Acetabularia. Science, N.Y.* **134**, 1361–1363.

Sweeney, B. M., Tuffli, C. F., and Rubin, R. H. (1967). The circadian rhythm in photosynthesis in *Acetabularia* in the presence of actinomycin D, puromycin and chloramphenicol. *J. gen. Physiol.* **50**, 647–659.

Vanden Driessche, T. (1966). Circadian rhythms in *Acetabularia*: Photosynthetic capacity and chloroplast shape. *Expl. Cell Res.* **42**, 18–30.

Vanden Driessche, T. (1966). The role of the nucleus in the circadian rhythms of *Acetabularia mediterranea. Biochim. biophys. Acta* **126**, 456–470.

Wilkins, M. B., and Holowinski, A. W. (1965). The occurrence of an endogenous circadian rhythm in a plant tissue culture. *Pl. Physiol. Lancaster* **40**, 907–909.

Author Index

Page, J. Ziegler, 69, 70, 71, *72*
Palmer, J. D., 34, *57, 58,* 62, 63, 64, 69, *72*
Papenfuss, H. D., 83, *92*
Papi, F., 61, *72*
Parker, M. W., 74, *90*
Pavlidis, T., 121, *135*
Penny, P. J., 99, 100, *102*
Pfeffer, W., 3, 5, 6, 7, 9, 10, *12,* 19
Pirson, A., 40, *57,* 88, *92,* 108, 109, 110, 114, *116*
Pittendrigh, C. S., 22, 24, 34, 37, 44, *55, 57,* 87, *92,* 121, *134*
Plinius, (Pliny), 1, *12*
Pogo, A. O., *116*
Pohl, R., 22, *57*
Prescott, D. M., 104, *116*
Pye, K., 97, *102*

Ralph, C. L., 65, *71*
Rao, K. P., 60, *72*
Rau, W., 78, *91*
Reimann, B. E., 104, *116*
Richter, G., 25, *57, 135*
Rieth, A., 105, *116*
Rogers, L. A., 18, 19, *57*
Rosensweig, N. S., 34, 37, *57*
Round, F. E., 34, *57, 58,* 62, 63, 64, *72*
Rubin, M. L., 34, 37, *57*
Rubin, R. H., 25, *58,* 130, *136*
Rückebeil, A., 128, *135*
Ruddat, M., 25, 26, 52, *58*
Ruge, A., 89, *92*

Sachs, J., 3, 5, 9, *12*
Sachs, R. M., 111, *116*
Sage, A. P., Jnr., 121, *135*
Salanki, A. S., 99, *102*
Salisbury, F. B., 75, 81, 82, 83, 85, *92*
Sargent, M. L., 34, 48, *58,* 101, *103*
Schleif, J., 66, *116*
Schmidle, A., 34, 37, *58*
Schoener, B., 96, *102,* 121, *134*
Schöne-Schneiderhöhn, G., 123, *134*
Schoser, G., 21, *57*
Schweiger, E., 25, 26, *58,* 122, *135*
Schweiger, H. G., 25, 26, *58,* 122, 130, *135, 136*
Scott, B. I. H., 96, *102, 103*
Semon, R., 7, 8, *12,* 40, *58*

Siegelman, H. W., 75, *90*
Smith, G. M., *72*
Sorokin, C., 108, 109, *116*
Spurný, M., 99, *103*
Stälfelt, M. G., 99, *103*
Stern, H., 110, *115, 116*
Stern, K., 11, *11*
Strumwasser, F., 122, *136*
Sussman, A. S., 34, *58*
Sweeney, B. M., 25, 28, 29, 31, 33, 35, 36, 37, 41, 43, 44, 45, 49, *56, 58,* 69, 70, 71, *72,* 112, *115, 116,* 122, 125, 128, 130, *135, 136*

Takimoto, A., 81, 84, 85, *91, 92*
Tamiya, H., 108, *116*
Tazawa, M., 21, *56*
Terborgh, J., 27, *57, 58*
Todt, D., 40, *58*
Toole, E. H., 74, *90*
Toole, V. K., 74, *90*
Trucco, E., 131, *135*
Tuffli, C. F., 25, *58,* 130, *136*
Tuttle, A. A., 99, 100, *102*

Uebelmesser, E. R., 34, 37, 41, *59*

Vaadia, T., 34, *59*
Vanden Driessche, T., 25, *59,* 124, *136*
Vielhaben, V., 53, 54, *56,* 65, 68, 69, *72*
Volcani, B. E., 104, *116*
Volm, M., 105, *116*
von Denffer, D., 105, 107, *115*

Wagner, R., 51, *59*
Walker, F. T., 101, *103*
Wallraff, H. G., 25, 26, *58,* 122, *135*
Warren, D. M., 22, 23, *59*
Wassermann, L., 40, *59*
Webb, H. M., 118, *134*
Wever, R., 16, 17, *17*
Wilkins, M. B., 22, 23, 38, 39, 52, *59, 136*
Woodward, D. O., 34, *58*

Yoshioka, T., 98, *102,* 121, *135*
Young, R. A., 101, *103*

Ziegler, J. R., *72*
Zimmer, R., 21, 44, 47, *56, 59,* 80, *92*
Zimmermann, W. F., 121, *135*

Subject Index

A

Acacia, leaf movements, 5, 6, 7, 8, 40
Acetabularia, photosynthesis, 25, 26, 122, 123, 130
  chloroplast shape, 26–27, 124, 130
  nucleus and rhythm, 113
Actinomycin D, 128, 129, 130, 131
Action spectra, 45, 46, 74–75
Active systems (*see also* Endogenous rhythms), 16
ADP, oscillations in concentration, 98
Albizzia julibrissin, leaf movements, 2
Algae (*see also under individual names*),
  blue-green, 18
  brown, 65
  filamentous, 18
  green, 40, 69, 89
  marine, 66
Allium cepa, nuclear size changes, 123
'Allowed time' for cell division, 112, 114
Alpha : rho ratio, (a/ρ), 17
Amethopterin, 129
Ammonium ions, rhythmic incorporation, 109
AMP, oscillations in concentration, 98
Amplitude of an oscillation, definition, 15
Anaerobiosis, rhythmic phenomena and, 53–54, 95
Anagallis, photoperiodism in, 78
Animal rhythms, 38, 44, 60–61, 66
Annual biological rhythms, 16, 73–90
Anthers, synchronous cell division in, 109–110
Aplysia, 122, 128
Arabidopsis, photoperiodism in, 78
Arhythmia, light induced, 38, 40
  possible causes, 39
Arsenate, effect on sporulation, 125
'Aschoff's rule', 38

Aspergillus ochraceus, growth rhythm, 101
Astasia, synchronization of cell division, 105
ATP, energy source for flagella, 93
  oscillations in concentration, 98
  plasmodial streaming rhythm and, 95
'Auto-phasing' of biological oscillations, 119, 120
Auxin level, rhythms of electrical potential and, 96
Avena, growth rhythm, 37, 38, 48, 54
Averrhoa bilimbi, short period oscillations, 98
Azide, rhythms and, 125

B

Bacteria, absence of circadian rhythms in, 18
Bamboo, flowering frequency, 93, 101
Barometric pressure, rhythm generation and, 118–119
Basic oscillator (*see also under* Oscillator), 45, 55, 111, 113, 122, 124, 128
'*Beats*', resulting from different rhythms, 67, 69
Biddulphia, synchronous cell division, 105
Biological clock(s) (*see also under* Clock), 34, 39
Biological rhythms (*see also under* Annual, Circadian, Diurnal, Lunar, Semilunar, Tidal),
  amplitude of, 15
  bioluminescence, (*see also* luminescence), 27, 28, 33, 133
  carbon dioxide evolution, 22, 23, 38–39, 48, 52
  carbon dioxide uptake, 25, 30, 52